"Creation is full of holy rhythms of life. There is a way to live that honors and embodies these rhythms. *The Common Rule* is a beautiful, inviting resource that helps us do just that. It is an important guide to living more deeply rooted in God's life-giving

Alan Fadling, president and founder, Unhurried Living

"One of the biggest problems among believers t Conse-quently, they do not live up to the expectations o can see in the decreasing morality in our country. *The Common Rule* presents a common-sense discipline that any follower of Jesus could use to become a joyful missionary disciple."

Michael Timmis, former chairman of Prison Fellowship Ministries, member of the board of Alpha International and The New Canaan Society

"Habitually choosing what is best over and above what is loud and urgent has never been more difficult than in a culture of perpetual distraction. 'But where do I begin?' people ask. In this book, Justin Earley offers the answer. Follow his lead, and you will find much of your life handed back to you."

John Stonestreet, president of The Chuck Colson Center for Christian Worldview

"It is encouraging to see a new generation of Christians wrestling with the formative power of habits and rituals. Our patterns of life shape our attitudes and actions more than we may realize. *The Common Rule* urges us to consider our habits in a new light and to embrace a new way of life in which we self-consciously limit ourselves in order to pursue what is best for us and for our neighbors."

Trevin Wax, director for Bibles and reference at LifeWay Christian Resources, author of *This Is Our Time: Everyday Myths in Light of the Gospel*

"Although I'm a church leader who has known the Lord for decades, I struggle to find space in my days to actually abide in Christ. Instead, I'm tethered to the demands that barrage me through my smartphone. *The Common Rule* offers a practical way back—not only for individuals but also for churches and small groups."

Karen Heetderks Strong, senior director of The Falls Church Anglican, Falls Church, Virginia

"Books are not meant to be just read. The good ones move you. The great ones change you. I'm now a fan of Justin Whitmel Earley. He's done something that is not so common: he teaches us eight habits that not only change our own lives but more importantly change the lives of those who choose to follow us."

Tommy Spaulding, author of *The Heart-Led Leader* and *It's Not Just Who You Know*

"Justin Earley offers a lifeline to every busy, smartphone addicted, distracted person on earth. In his deeply personal and immensely practical book, he inspires us to find a rhythm that will support our most life-giving relationship of all—our friendship with Jesus."

Ken Shigematsu, pastor of Tenth Church, Vancouver, BC, author of *Survival Guide for the Soul*

"In the spirit of Richard Foster, Eugene Peterson, and so many other reflective authors, this book on the common rule is an exciting contribution to the family God. Each of us needs and longs for a path to deepen our sense of being in this world. So many Christian writers just give us one more self-help book. *The Common Rule* breaks that mold and serves as a call to deeper intimacy with God."

Gary Bradley, The Navigators

"Justin Whitmel Earley's honest walk through anxiety gives hope to anyone who needs a new way. *The Common Rule* becomes an accessible way to move forward in your walk with Christ, whether battling anxiety or simply wanting to find Christ at the center of your life. Earley's fresh look at a liturgical life comes off the page and into your daily life. This book is a refreshing reminder that community, shared meals, fasting, praying, silence, and rest all have deep effects on our lives. In a society of fast food and split families, the return to the table is more important than ever. Earley lived the fast-paced world of no margin where everyone has free access to his life, yet in the pages of this book, we see he has found another way–another way for us all."

Diana M. Shiflett, pastor of spiritual formation, Naperville Covenant Church, author of *Spiritual Practices in Community*

"When someone asks how you're doing and you always find yourself answering, 'So busy and crazy,' it might be time for a change. *The Common Rule* offers practical wisdom on how to slowly but deliberately restructure our lives, and it shows us that when we embrace limitations, we paradoxically gain the freedom we long for."

John Dyer, author of *From the Garden to the City: The Redeeming and Corrupting Power of Technology*

"I'm thankful for *The Common Rule* because it is a practical tool to save me from the tyranny of machine-like productivity. The book and practices remind me that I am a human being and not a human doing. Read this book and save yourself from the tyranny!"

David M. Bailey, executive director of Arrabon, coauthor of *Race, Class, and the Kingdom of God*

"In the present age, we are settling to be informed when the call is to be transformed. I'm no Luddite, but we must rethink the way technology and busyness are affecting our loves. *The Common Rule* not only has incredible practical advice for daily and weekly rhythms, it also opens our eyes to see the water we're swimming in. Every church should consider using this as a resource for formation."

AJ Sherrill, author of *Enneagram and the Way of Jesus*, lead pastor at Mars Hill Bible Church, Grand Rapids, Michigan

"This is a wonderfully practical book. But even more, it is a beautiful one—full of glimpses of the life we were made for and the simple choices that can take us in that direction. The habits described here and the wisdom they embody are the path toward sanity in a frenetic world."

Andy Crouch, author of *The Tech-Wise Family* and *Culture Making*

"*The Common Rule* is an engaging, relevant, and transforming guide for cultivating spiritual discipline amid the distractions and attractions of life and the daily pressures and challenges of just getting by. The spiritual practices Justin offers are not simply habits and forms adapted for modern life but are habits and forms developed in the painful crucible of personal experience. This book is worth reading and its ways worth pursuing."

Ron Nikkel, president emeritus, Prison Fellowship International

"I love the practicality of this book. Justin Whitmel Earley understands and embodies the reality that we show what we love and value by the daily habits of our lives."

Mark Scandrette, author of *Free* and *Practicing the Way of Jesus*, and coauthor of *Belonging and Becoming*

JUSTIN WHITMEL EARLEY

THE COMMON RULE

HABITS OF PURPOSE FOR AN AGE OF DISTRACTION

An imprint of InterVarsity Press
Downers Grove, Illinois

InterVarsity Press
P.O. Box 1400, Downers Grove, IL 60515-1426
ivpress.com
email@ivpress.com

InterVarsity Press® is the book-publishing division of InterVarsity Christian Fellowship/USA®, a movement of students and faculty active on campus at hundreds of universities, colleges, and schools of nursing in the United States of America, and a member movement of the International Fellowship of Evangelical Students. For information about local and regional activities, visit intervarsity.org.

Author photo: Franklin Earley
Cover design: David Fassett
Interior design: Daniel van Loon
Images: branch with leaves: © C Squared Studios / Photodisc / Getty Images
textured background: © Pinghung Chen / EyeEm / Getty Images

ISBN 978-0-8308-4560-6 (print)
ISBN 978-0-8308-7338-8 (digital)

Printed in the United States of America ♾

InterVarsity Press is committed to ecological stewardship and to the conservation of natural resources in all our operations. This book was printed using sustainably sourced paper.

Library of Congress Cataloging-in-Publication Data

Names: Earley, Justin Whitmel, 1984- author.
Title: The common rule : habits of purpose for an age of distraction / Justin Whitmel Earley.
Description: Downers Grove : InterVarsity Press, 2019. | Includes bibliographical references.
Identifiers: LCCN 2018050387 (print) | LCCN 2019001130 (ebook) | ISBN 9780830873388 (eBook) | ISBN 9780830845606 (pbk. : alk. paper)
Subjects: LCSH: Christian life. | Spiritual life.
Classification: LCC BV4501.3 (ebook) | LCC BV4501.3.E2655 2019 (print) | DDC 248.4/6—dc23
LC record available at https://lccn.loc.gov/2018050387

P 26 25 24 23 22 21 20 19 18 17 16 15 14 13 12 11
Y 38 37 36 35 34 33 32 31 30 29 28 27 26 25 24 23 22 21

TO LAUREN

"In binding love you set me free."

CONTENTS

RESOURCES

DISCOVERING THE FREEDOM OF LIMITATIONS

It was twelve on an ordinary Saturday night when I woke suddenly in a dreadful panic, sweating and shaking. I sat up in bed in the silence of the bedroom that my wife, Lauren, and I share. The feeling was so intense I expected to find something terrible had happened, as if my subconscious knew something I didn't. But all was quiet.

It was such an odd moment that I woke Lauren and tried to explain it to her, but there wasn't much to explain. It was like my heart had rung the alarm bells for no reason. Finally, after a while of trying to calm down, I fell back asleep.

The next day, a vague feeling that something was wrong lingered. That afternoon we took our sons apple picking in the mountains west of our city, Richmond, Virginia. The orchards had the stunning beauty of late September. We ate apple cider donuts, and my wife and I watched the boys run around the trees. It should have been perfect—but I was only half there. It was like my emotions were wearing sunglasses; everything had a shade of nervousness.

That night, the same thing happened, but this time I never fell back asleep. I spent Monday at the office like a zombie, hunched over papers and traveling back and forth from my desk to the terrible single-serve coffee machine. The fear was starting to work through me like a virus. I dreaded the moment when I would have to lie back down that night with my panicked thoughts.

When I did, it all started again.

So it was that I ended up in the emergency room at three in the morning, looking at a doctor who half-apologetically told me nothing was wrong. I was just showing symptoms of clinical anxiety and panic attacks. He assured me—as if it were comforting—that these were very common. I couldn't believe what I was hearing.

EVERYTHING IS FINE, EVERYTHING IS FALLING APART

I couldn't believe it because as far as I knew, I wasn't stressed or worried about anything. Actually, everything seemed to be going really well. After studying English literature at the University of Virginia and marrying my wonderful wife, we spent a few years living in China as missionaries. I loved living there, and we would have stayed even longer, except that one day I saw something that changed the way I viewed the world. I was taking a walk on a pedestrian street, and in the space of ten minutes, I came across someone dealing drugs, someone running a brothel, someone selling stolen laptops, and someone protesting the government.

Except for the political protest, all of those were normal in China. In my four years there, I had never seen a protest, and I would never see one again. I watched as she unfurled a sign that said, "The judicial system in China is broken, the people in the countryside are being oppressed"—she was arrested so quickly that I never read the rest of it.

As I walked away, I reflected on how all four of those things were illegal, but three of them were considered to be legitimate ways to make money. And out of the four, only one was a brave act of love for neighbor—punishable by arrest.

That was the day I realized the power that law and business have in shaping the world, and I felt a tremendous sense of calling. I felt the Lord telling me that if I wanted to follow him, I should do it in those arenas. That's where he wanted me to be a missionary. And I listened. So Lauren and I moved to Washington, DC, where I went to Georgetown Law and Lauren began her career in philanthropy consulting.

During this time, our oldest two sons, Whit and Asher, were born. After graduating at the top of my class at Georgetown, I landed a job as a mergers and acquisitions attorney at the best big law firm in Richmond. All my best friends and family lived in Richmond, so we moved down there to live happily ever after—or so I thought.

I was thrilled with life that summer. I grew an enormous beard (convincing my wife that I would have to shave it as soon as I started at the law firm), bought a vintage BMW motorcycle (convincing my wife that it would be a convenient mode of transport), and spent all my time when I wasn't studying for the bar exam either working out or playing with my sons (Lauren needed no convincing on this one).

In short, life was going great—except for one thing. I was tired. Really tired. In my years since graduating from college, I had tackled all of life with a voracious hunger. I wanted to be good at everything I did. I spent my years in China up early studying Mandarin and up late hanging out with fellow missionaries and Chinese friends.

In law school, my life became an endless series of calendar alerts, appointments, résumé-building activities, and studying late into the night. But we were all like that, so nothing about it seemed strange. I remember at the bar exam telling my friends I'd had trouble sleeping the night before the test, and they all looked at me strangely; apparently I was the only one who hadn't brought sleeping pills.

Being overwhelmed by ambition was a way of life in law school, so I went along with it. I thought that was how you got to be a top law student, got the big firm jobs, and became a successful young lawyer—by saying yes to everything and no to nothing.

So I was way too busy, totally overcommitted, and living with a chaotic, packed schedule. But I thought I was different because *I had a calling*. After I saw that protester arrested, I had become consumed with the idea of how important law and economics are in shaping the culture we live in—for better or for worse.

Only in retrospect did I realize that, while the house of my life was decorated with Christian content, the architecture of my habits was just like everyone else's. And that life had been working for me—until it collapsed.

THE MISSIONARY GETS CONVERTED

That night in the hospital, the doctor gave me a bottle of sleeping pills and told me I needed to slow down. Of course I had no idea how to slow down. Busyness functions like an addiction. When you stop, you have to face your thoughts, which terrifies most of us. So to cope you have to keep up the busyness. I started taking the sleeping pills and thus began the darkest phase I have ever known.

The pills blacked me out for a couple of hours every night, but I soon discovered that I react to sleeping pills in all the horrible ways you read on the back of the bottle. Enormous mood swings during the day, hallucinogenic nightmares, even suicidal thoughts—those were all me. In a matter of weeks, I had become disturbingly unstable, crying at random, struggling to focus my thoughts, and unable to calm my groundless fears.

I was standing in the kitchen one night, and Lauren handed me some dishes to put away. I looked at her and said, "I don't know where these go." And I really didn't. My mind was getting so thin that simple tasks were becoming overwhelming. *If I can't put away dishes,* I thought, *how am I going to be a husband, a friend, and a father?* Everything I held dear was being threatened.

This began a long stretch of time when I needed either pills or alcohol to fall asleep. Eventually I ditched the sleeping pills—thank God—but the drinks remained. I needed a few to fall asleep. My conversion from the young missionary to the medicating lawyer was now complete. And a stubborn question appeared: *How did the missionary come to be the one that got converted?*

Answering this question was not easy at all. It was a very long and difficult road. I now see that my body had finally become converted to the anxiety and busyness I'd worshiped through my habits and routines. All the years of a schedule built on going nonstop to try to earn my place in the world had finally rubbed off on my heart. My head said one thing, that God loves me no matter what I do, but my habits said another, that I'd better keep striving in order to stay loved.

In the end, I started to believe my habits—mind, body, and soul.

That's why what happened next was so important. Fifteen months later, I came to another evening that changed my life.

DISCOVERING THE POWER OF HABIT

It was an evening just after New Year's Eve, and I was sitting in a restaurant with my two best friends, Matt and Steve. On the table was a paper covered with scribblings: a program of habits my wife and I had sketched out. The aim of the program was to try to get my heart to believe the peace that my head professed but my body refused. Matt and Steve were going to keep me accountable.

It's important to know just how unremarkable that evening was at the time. There was no great revelation, no victory. It was just a night of good conversation about living with better daily and weekly rhythms. I didn't think any of the habits that were written out at that table were going to be *that* important. There were daily habits of prayer and some hours away from my phone, and weekly habits of taking a day off work and keeping up conversations with friends. But none of them stuck out as a new discovery.

At the time, I had never heard of a keystone habit—a micro shift that brings about macro effects. I didn't think that a few habit shifts would change my life, but I was willing to try anything. Today, as I write these words, I still live according to those habits, I still work the same job, I still have the same calling, but I slept like a baby last night. I slowly became a new man—a humbled man, but a much stronger man.

I didn't think these habits would matter much because I had no idea how much my ordinary habits were shaping my soul in the most extraordinary ways. I had no idea how much my life was being formed by my habits instead of my hopes. Most of us don't, of course, because habits are the water we swim in.

SEEING THE WATER

On May 21, 2005, David Foster Wallace gave his electric and instantly famous graduation address at Kenyon College, called "This Is Water." It began with this narrative:

> There are these two young fish swimming along, and they happen to meet an older fish swimming the other way, who nods at them and says, "Morning, boys, how's the water?" And the two young fish swim on for a bit, and then

eventually one of them looks over at the other and goes, "What the hell is water?"

The point is, as he put it, "the most obvious, important realities are often the ones that are hardest to see and talk about."[1] As far as our habits go, the invisible reality is this: *We are all living according to a specific regimen of habits, and those habits shape most of our life.*

A habit is a behavior that occurs automatically, over and over, and often unconsciously. A study from Duke University suggested that as much as 40 percent of the actions we take every day are not the products of choices but of habits.[2] As William James put it, "All our life, so far as it has definite form, is but a mass of habits."[3] The problem is, as Wallace suggested, that much of what is fundamentally shaping our existence is happening unconsciously.

But just because we don't choose our habits doesn't mean we don't have them. On the contrary, it usually means someone else chose them for us, and usually that someone doesn't have our best interests in mind.

Take your work schedule or your social media scrolling, for example. Think about your internet history or how you spent your mornings last week. Think about what you usually eat for lunch or the time you spend with family versus the time you spend looking at a screen during a normal day. These things define vast portions of our lives, and while we would like to think we've carefully chosen them, most often we haven't even given them a second thought. Most often we just swim along with what those around us are doing. And much more often than we would like to admit or even understand, we are nudged into those choices by those who want to make money off the patterns of our daily life.

This wouldn't be so bad if it weren't for the fact that habits form much more than our schedules: *they form our hearts.*

THE SCIENCE AND THEOLOGY OF HABIT

In his compelling book *The Power of Habit*, Charles Duhigg wrote that "when a habit is formed, the brain stops fully participating in decision making. The patterns we have unfold automatically."[4] Brain activity during habits happens in the deepest part of the brain, the basal ganglia. This saves us a lot of mental energy for other thoughts. So we can get in the car and suddenly arrive home without thinking about a single turn we made. Instead we've been thinking over a sticky work problem or an ill relative. Habits allow us to put our brains to better use.

This is really useful in general, but it has downsides. First, if we're acting out a bad habit—one that reinforces an addiction, perpetuates a harmful pattern of thought, or encourages mindless submission to a technology that is designed to attract our attention and sell it to the advertisers—we don't have much power to fight back. We may know that something is unhealthy or wrong. We may know exactly why it's bad or undesirable. We can tell ourselves that over and over, but that part of our brain is exactly the part that gets shut out when the autopilot of habit turns on.

Second, because our unconscious choices form us just as much, if not more than, our conscious ones, we can become formed in patterns that we would never consciously choose if we were aware of them. This is the difference between what we call education and formation. Education is what you learn and know—things you are taught. Formation is what you practice and do—things that are caught. The most important things in life, of course, are caught, not taught, and formation is largely about all the unseen habits.

This is why to fully understand habits you must think of habits as liturgies. A liturgy is a pattern of words or actions repeated regularly as a way of worship. The goal of a liturgy is for the participant to be formed in a certain way. For example, I say the

Lord's Prayer every night with my sons because I want the words of Jesus' prayer to sink down into their bones. I want that prayer to form the contours of their lives.

Notice how similar the definition of liturgy is to the definition of habit. They're both something repeated over and over, which forms you; the only difference is that a liturgy *admits* that it's an act of worship. Calling habits liturgies may seem odd, but we need language to emphasize the non-neutrality of our day-to-day routines. Our habits often obscure what we're really worshiping, but that doesn't mean we're not worshiping something. The question is, what are we worshiping?

As philosopher James K. A. Smith argues in his book *You Are What You Love: The Spiritual Power of Habit*, the habits we play out day after day are not tangential to our worship but actually central to it. Worship is formation, and formation is worship. As the psalmist put it, those who make and trust in idols will become like them (Psalm 31:6). So we become our habits.

When we combine Smith's insight that our habits are liturgies of worship and Duhigg's neurological insight that our brains aren't totally engaged when our habits are playing out, we have a robust explanation of how our unconscious habits fundamentally reshape our hearts, regardless of what we tell ourselves we believe.

SMALL HABITS AS POWERFUL LITURGIES

To make this concrete, let me show you how this played out in my daily routine before my anxiety crash.

Table 1. Bad Habits as Liturgies of Wrong Belief

Habit	Liturgy of Wrong Belief
Wake up exhausted again, because I never get to bed on time.	I am not a creature; I am infinite. My body will be fine. I am a god.

Habit	Liturgy of Wrong Belief
Look at work emails on my phone before getting out of bed.	I can miss a quiet time, but I can't miss a quick response. Unless I'm well regarded in the office, I'm not worth anything.
Grab breakfast on the go, while everyone else in my family scrambles to get somewhere late. At the office, eat lunch at my desk.	Being too busy is normal, and maybe even desirable. I'm important if a lot of people want my time. To stay important, I need to stay busy, and that means being late all the time.
Keep all computer notifications turned on, and keep my phone on and in sight while I work.	I need to know what's going on out there. The most recent thing is the most important thing. The best way to love my neighbors is to stay updated on dramatic headlines and new memes, not to do focused work.
If a manager asks for something late in the day on an unrealistic deadline, always say yes. If a social invite comes up, always go for it.	I will become the best version of myself by expanding my options, so I can't say no. I may be tired and busy, my family may be exhausted by my unpredictability, but if I don't preserve choice, I can't be who I really am.
Even when I sense all of the above is getting out of control, even when the best word to describe life is "scattered" or "busy," resist any rules that would restrict technology use and work schedules.	To limit myself is to restrict my freedom. And I'm not fully human without my freedom of choice in every moment. The good life comes from choosing what you want.

We can stop there. I'm not even halfway through my day, and you can see how by *not* having a program of habits I was submitting to a rigorous regimen of liturgies simply by assimilating to the usual way of life in America. My life was an ode of worship to omniscience, omnipresence, and limitlessness. *No wonder my body rebelled.*

THE SLAVERY OF FREEDOM

All of these liturgies of wrong belief were important in forming me to be anxious, but the last one on the chart is especially dangerous:

the freedom liturgy. Why is the freedom liturgy so dangerous? Because it perpetuates the slavery to all the other habits—ironically.

The freedom liturgy is dangerous for two reasons. First, it doesn't actually produce freedom. We think that by rejecting any limits on our habits, we remain free to choose. Actually, by barraging ourselves with so many choices, we get so decision-fatigued that we're unable to choose anything well. Since we're too tired to make any good decisions, we're extremely susceptible to letting other people—from manipulative bosses to invisible smartphone programmers—make our decisions for us. The dogged pursuit of this kind of freedom always collapses into slavery, which leads us to the second reason the freedom liturgy is dangerous.

MY LIFE WAS AN ODE OF WORSHIP TO OMNISCIENCE, OMNIPRESENCE, AND LIMITLESSNESS. *NO WONDER MY BODY REBELLED.*

The second reason is that it blinds us to what the good life really is. When we act out the "no-limits-none-ever" freedom liturgy, we assume that the good life comes from having the freedom to do whatever we want. So to ensure the good life, we have to ensure our ability to choose in each moment. *But what if the good life doesn't come from having the ability to do what we want but from having the ability to do what we were made for? What if true freedom comes from choosing the right limitations, not avoiding all limitations?*

In retrospect, that night in the restaurant with my friends, sketching out a habit plan, was such a big moment because I finally surrendered the keystone habit of freedom. I decided limits were a better way of life, *and that's when everything changed*. I had lived my whole life thinking that all limits *ruin* freedom, when all along it's been the opposite: the right limits *create* freedom.

This was no overnight realization, but as my life began to change, I began to wonder why surrendering the freedom liturgy and accepting limitations was so hard for me and for Americans generally. I began to wonder how we had come to believe such a bizarre definition of freedom, and whether there were living examples of a better freedom.

I found the answer in the life of Jesus.

JESUS AS THE GOOD MASTER

There's no one who surrendered more freedom than Jesus, who went from the all-powerful second person of the Trinity to the vulnerable form of a helpless infant. He went from speaking the universe into existence by his Word to not being able to speak a word. This is what the Scriptures mean when they say that he "emptied himself" (Philippians 2:7).

But it doesn't stop there; he did not just become human. He became a poor human. A homeless human. A human who loved with such power that he became a threat to those in power, so they tortured and killed him. Jesus submitted to the ultimate limitation: to be snuffed out of the world in death. But why? Why would he do this?

For love.

For love of you and me.

Philippians says that because he was willing to submit to the limitations of death, he was exalted. When Jesus got up and walked out of the grave, he exploded the limitations of what it meant to be human by dancing on death itself. Now those who choose to surrender their life to Christ will also rise with Christ. By surrendering his freedom for the sake of love, Christ saved the world. By surrendering our freedom to him, we participate in that love. We find our true freedom in the constraints of divine love.

The key thing to notice here is how Jesus' actions are the exact opposite of what humans did in the Garden of Eden. There, we tried to become gods by rejecting God's authority and eating the forbidden fruit. In trying to free ourselves from our limitations, we brought the ultimate limitation of death into the world. But Christ turns this human paradigm on its head. The way down is the way up. The way to victory is through surrender. The way to freedom is through submission.

We, for our own sake, tried to become limitless, and the world was ruined. Jesus, for our sake, became limited *and the world was saved.*

WE, FOR OUR OWN SAKE, TRIED TO BECOME LIMITLESS, AND THE WORLD WAS RUINED. JESUS, FOR OUR SAKE, BECAME LIMITED *AND THE WORLD WAS SAVED.*

DISCOVERING THE RULE OF LIFE

This was all new to me, but of course it isn't new at all.

As my personal life was changing, so was my work life. Surprisingly, by putting limits on my work schedule and technology use, I found I became better at my job. These new habits helped me to focus much more on work. After putting up some boundaries around my time, I found that people actually "needed" me a lot less than I thought. One of my habits was to turn my phone off for one hour each evening, and I found that most of my clients and colleagues were fine getting a call back an hour later in the evening when I turned my phone back on.

Consequently, I started talking about habits a lot. I probably annoyed a lot of my friends, who had to hear about it over and over. One day I was explaining some of my realizations about habits to my pastor, and he asked to see what I was doing. I'll never forget what he said when he looked at them: "Oh, I see. You've crafted your own rule of life."

I answered with the ironic words of someone who would go on to write a book about it: "What's a rule of life?"

I now know that a "rule of life" is a term for a pattern of communal habits for formation. The most well-known rules of life were originally developed by church fathers and ancient monastics, such as St. Augustine or St. Benedict. But for thousands of years, spiritual communities have been using the frame of the rule of life as a mechanism of communal formation.

Despite our understanding of the word "rule," a "rule of life" is much less about obeying rules than it is about finding communal purpose. For example, while both St. Augustine's and St. Benedict's rule have all kinds of tiny habits that we might either consider too inane to matter or too strict to be appropriate, we should notice that both of them had the same goal in mind: *love*.

Both were obsessed with taking the small patterns of life and organizing them towards the big goal of life: to love God and neighbor. St. Augustine's rule began with this sentence: "Before all things, most dear brothers, we must love God and after Him our neighbor; for these are the principal commands which have been given to us." St. Benedict's rule opens declaring that it means to establish "nothing harsh, nothing burdensome," but goes on to describe walking in God's commandments as being in the "ineffable sweetness of love."

Both saw habits as the gears by which to direct life toward the purpose of love. In fact, the word *rule* is used because it comes from the Latin word *regula*, a word associated with a bar or trellis, the woodwork on which a plant grows. The idea is that we (like plants) are always growing and changing. But when there is no order, growth can take something that was supposed to produce fruit and turn it into a twisted vine of decay. That description was frighteningly accurate in my case. The rule of life is intended to pattern communal life in the direction of purpose and love instead of chaos and decay.

COMMUNAL HABITS AS A CONTEMPORARY RULE OF LIFE

Anyone who stops to think about it comes to the same realization—humans are largely defined by the small routines that make up our days and weeks. Author Annie Dillard wrote, "How we spend our days is, of course, how we spend our lives. What we do with this hour, and that one, is what we are doing. A schedule defends from chaos and whim. It is a net for catching days. It is a scaffolding on which a worker can stand and labor with both hands at sections of time."[5]

ONLY WHEN YOUR HABITS ARE CONSTRUCTED TO MATCH YOUR WORLDVIEW DO YOU BECOME SOMEONE WHO DOESN'T JUST KNOW ABOUT GOD AND NEIGHBOR BUT SOMEONE WHO *ACTUALLY LOVES GOD AND NEIGHBOR.*

The best way to understand a "rule of life" or a program of habits is by picturing Dillard's scaffolding, her version of the trellis. Habits are how we stand up and get our hands on time. And because time is the currency of our purpose, *habits are how we get our hands on our purpose*. If you want to get your hands on what you know, you need to seek out the right words. If you want to get your hands on who you are becoming, you need to get your hands on habits. A rule of life is how we get our hands on our habits. Understanding this insight, monastic orders have for millennia lived according to rules of life for the purpose of unifying head and habit.

It's high time that this ancient spiritual wisdom become modern common sense. All those who want to be attentive to who they are becoming *must* realize that formation begins with a framework of habits.

It's utterly important to learn the right theological truths about God and neighbor, but it's equally necessary to put that theology into practice via a rule of life. You can't believe truth without practicing truth, and vice versa. You can't have a good education without good formation, and vice versa. You can't know who Jesus is without following Jesus, and vice versa. To live with only one or the other of these things is to live as a half-human being.

Only when your habits are constructed to match your worldview do you become someone who doesn't just know about God and neighbor but someone who *actually loves God and neighbor.*

THE COMMON RULE

As it turns out, I had stumbled upon some ancient wisdom. My little program of habits was a sort of rule of life applied specifically to the deformed modern liturgies of business and technology.

As I talked with my friends and family about how living according to a rule of life was changing my life, many of them suggested I share it so others could try it. So I took some of my favorite habits, put them together in a PDF, and called it "The Common Rule" because it was intended for *common* practice by *common* people. Then I emailed it out to about fifteen friends.

Within a week, it was forwarded to hundreds of people. From there it kept going. I had no idea how normal I was. I had no idea how many other people were also desperate for a more sensible and meaningful way to organize their days and weeks.

Since I began writing about the Common Rule, I've learned that the vast majority of contemporary Americans I encounter are—just like me—absolutely starved for an example of how to order daily life in a way that unites head and habit.

However, as I talked with more people and read more books on habits, formation, and liturgy, my main concern changed. I realized the most alarming part of this is not our bad habits, which we tend to know about. It's our collective assimilation, which is invisible to us.

We have a common problem. By ignoring the ways habits shape us, we've assimilated to a hidden rule of life: the American rule of life. This rigorous program of habits forms us in all the anxiety, depression, consumerism, injustice, and vanity that are so typical in the contemporary American life.

It's urgent, then, that we recover the wisdom of crafting a gospel-based rule of life as the new norm for living as a Christian in America today. We desperately need a set of counter-formative practices to become the lovers of God and neighbor we were created to be.

This is not just a personal matter. It is a public matter of neighbor love. Talking about Jesus while ignoring the way of Jesus has created an American Christianity that is far more American than it is Christian. Paying all our spiritual attention to the message of Jesus while ignoring his practices has not only led people like me into devastating life crises, it has also created a country of Christians whose practical lives are divorced from their actual faith. How else do we explain a country of Christians who preach a radical gospel of Jesus while assimilating to the usual contours of American life?

There is a better way. It is the way of Jesus.

Let us see that habits shape the heart. Let us stop fearing that limits are a threat to our freedom. Let us see that the right limitations are the way to the good life. Let us build a trellis for love to grow on. Let us craft a common rule of life for our time, one that will unite our heads and our habits, growing us into the lovers of God and neighbor we were created to be.

PART ONE

HOW TO PRACTICE THE COMMON RULE

WHAT IS A RULE?

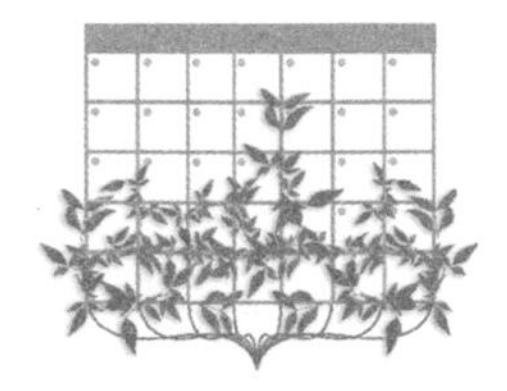

A "rule" is a set of habits you commit to in order to grow in your love of God and neighbor. As you may expect, this book is meant to be practiced, not simply read. Consider this book a companion, a handbook for trying out the Common Rule.

As the name suggests, the Common Rule is meant to establish communal—not individual—rhythms, so ideally it's practiced with other people. Change—even personal change—almost always happens within a community where people support each other, process what they're learning, and keep each other accountable to goals. So I strongly encourage you to convince friends or family members to try the Common Rule out with you.

Here are some ways you can do that.

TIMEFRAMES FOR PRACTICING THE COMMON RULE

Studies show that it takes at least two to three weeks to turn new practices into habits.

For a month. Committing to practice the Common Rule alongside other people for one month is the best way to indwell the rhythms of the habits and allow them to replace the unseen habits you never knew you had. This gives you the truest experience of the habits and gives you the best chance of letting the habits stick. (See "Trying a Month of the Common Rule" in Resources for information on how to plan your month.)

A month also allows you to get over the adjustment phase. Any new pattern of habits seems overwhelming until you realize there's a groove to it. The Common Rule will look like a lot until you realize that the rhythms lighten your load, they don't add to it. This usually takes a few weeks to sink in. (If you're still concerned, see the section that follows, "The Light Burden.")

For a week. If you aren't quite ready for a month, try a week. This book has eight chapters that focus on daily and weekly habits. Read one chapter each day (each is about a fifteen-minute read), and you'll finish one week later, on the same day of the week you started. If you're reading this with a group of friends or a small group you meet with weekly, try starting it on the morning before you meet for the first week, and you'll finish on the morning before you meet the second week.

Don't worry about trying to do all the habits every day. Just experiment with each habit on the day you read about it. From there, you can decide whether to try a month of the Common Rule. (See "Trying a Week of the Common Rule" in Resources for information.)

For a season. Practicing the Common Rule habits can be a great way to lean into a season of the year. For example, consider using the practices at the turn of a new year in order to make habits that stick instead of making resolutions you forget. You also may consider getting your spiritual community to practice the habits through a liturgical season, such as Lent or Advent. (See the resources on thecommonrule.org for some seasonal variations on the practices.)

I generally live according to these practices, but I often get even more serious about them or have someone keep me accountable when I'm going through a stressful season in my work or personal life. For example, while writing this book with a full-time job, three little boys in the house, and a baby on the way, the Common Rule habits helped me maintain a relatively meditative, rested, and focused background rhythm, despite that very full and potentially exhausting season of life. The year I spent writing this book was difficult, but practicing the Common Rule while writing about it kept a hard season from being a dangerous season.

So don't be afraid to start the Common Rule during a difficult season. In fact, it's designed to help you navigate difficult seasons by pointing you toward the love of God and neighbor, which is the path to follow in any season. It may be just the thing you need.

Try one or two. If you think you'd rather just browse the book and try out a habit or two, that's wonderful. There's no reason you can't skip around and read about whichever habit interests you. Each chapter stands on its own, so read about one that interests you, and consider trying it out. Many people find that the most impactful daily habit is Scripture before phone, while the most impactful weekly habit is Sabbath. (See "Trying One Habit of the Common Rule" in Resources for information on how to try one habit at a time.)

THE LIGHT BURDEN

One of the biggest (and most understandable) misconceptions people have about the Common Rule is that it will be a lot of work and take a lot of time. Don't worry, it won't. In fact, if you're feeling overwhelmed, if you feel like you're struggling to figure out how to make life work from day to day, you're holding the right book.

Let me tell you what *is* overwhelming: a default, normal, unexamined American life. *That* is completely overwhelming. It's so much to take on, and we all do it simply by not doing anything else instead.

The Common Rule is a different way to live. It's meant to distill your habits, so you do more meaningful things by doing fewer things. So whether you are in a season of moving, having children, caring for an aging parent, transitioning in your career, processing the loss of a loved one, or taking on a major new project or client, the Common Rule practices can turn the volume of life down to an appropriate level and focus you on loving God and neighbor.

It's challenging to make new habits. So I won't tell you the Common Rule is not hard. It is hard, but that doesn't tell you much. Anything worth doing is hard. What I'm trying to tell you is that it is *freeing*.

You'll find that once new Common Rule habits are established, by definition they don't take up time and mental space. They work in the background. They're designed to free up your time, create meaningful space for relationships, turn your energy toward good work, and focus your presence on the God who made you and loves you. That is not constricting; that is liberating. You were made for it.

HABITS AT A GLANCE

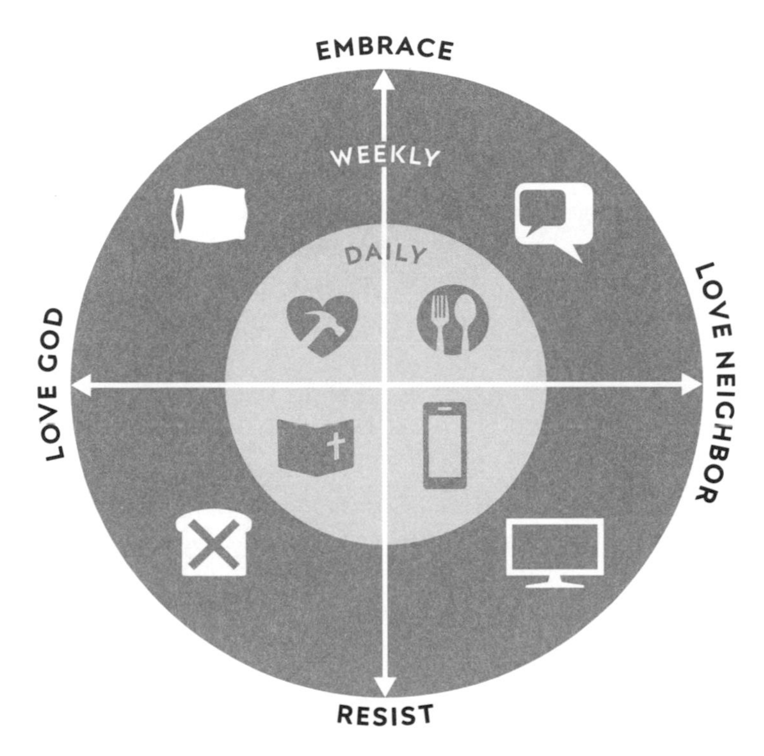

DAILY HABITS	WEEKLY HABITS
1 Kneeling prayer three times a day	1 One hour of conversation with a friend
2 One meal with others	2 Curate media to four hours
3 One hour with phone off	3 Fast from something for twenty-four hours
4 Scripture before phone	4 Sabbath

THE EIGHT HABITS OF THE COMMON RULE

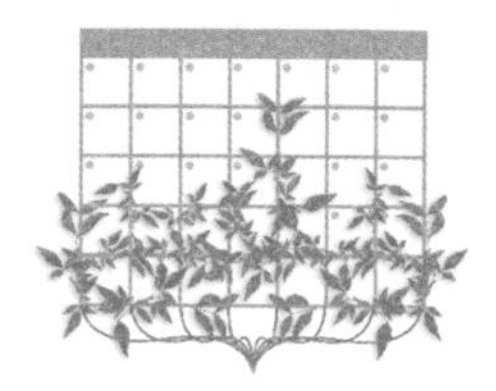

The Common Rule is made up of eight habits, four daily and four weekly.

The daily habits are

- kneeling prayer at morning, midday, and bedtime,
- one meal with others,
- one hour with phone off, and
- Scripture before phone.

The weekly habits are

- one hour of conversation with a friend,
- curate media to four hours,
- fast from something for twenty-four hours, and
- sabbath.

Each habit also corresponds to two different spectrums. The first spectrum is the love of God and the love of neighbor, with four habits focused on each. The second spectrum contrasts embrace and resistance, with four habits designed for each.

Love of God. Another way to look at the habits is as they pertain to love of God. You were made to love and be loved by God. Only in the light of his love will you finally see who you really are, feel how you are supposed to feel, and discover what you should do with your days. Thus four of the habits of the Common Rule are pointed toward opening our eyes to who God is, accepting the love he freely offers, and returning the gaze that has always been fixed on us:

- sabbath,
- fasting,
- prayer, and
- Scripture before phone.

Love of neighbor. When we think of better habits, we often think about our own self-improvement. Nothing could be further from the purpose of the Common Rule. These habits are meant to be practiced with others and for their sake:

- meals,
- conversation,
- phone off, and
- curate media.

The word *neighbor* here is used in the New Testament sense of the word. A neighbor is anyone and everyone who needs our love: family, friends, strangers, and enemies. These four habits mean spending meaningful time with other people. They encourage us to interrupt our busy schedules for the sake of rhythms of community. They encourage us to put down our devices and become more present with others. A friend asked me whether the Common Rule helped us care for ourselves, and my answer was, "yes, because we're made to be happy when we're focusing on others." These habits are designed to help us spend our days for the sake of others, rather than just ourselves.

Embrace. Embrace is a reminder that there is much good in the world God made. God's presence—not his absence—is the primary fact of the world. That we need each other—not that we harm each other—is the primary truth of being human. In the habits of embrace, we try to train our bodies and our hearts to love God as he actually is and to turn to our neighbor as we were made to do. The habits of embrace are

- sabbath,
- prayer,
- meals, and
- conversation.

Resistance. When we practice resistance, we acknowledge that evil and suffering are very real, though they aren't how the world was made to be. Our world is full of a thousand invisible habits of fear, anger, anxiety, and envy that we unconsciously and consciously adopt. Should we do nothing, we *will* be taught to love the very things that tear us apart. So we must take up the fight, open our eyes to the way media form us in fear and hate, the way screens form us in absence, and see the way excess and laziness train us to love ourselves above all else.

But remember that resistance has a purpose: *love*. The habits of resistance aren't supposed to shield you from the world but to turn you toward it. They aren't so you can feel good about what *you've* done for *you*. They exist so you can feel peace about what God has done for you. The habits of resistance are

- fasting,
- Scripture before phone,
- phone off, and
- curate media.

PART TWO

THE DAILY AND WEEKLY HABITS

KNEELING PRAYER AT MORNING, MIDDAY, AND BEDTIME

"Your kingdom come, your will be done, on earth as it is in heaven."
JESUS (MATTHEW 6:10)

Ora et labora. Work and pray.
MOTTO OF THE BENEDICTINE MONKS

BEING A CORPORATE LAWYER

When people ask me what I do and I don't want to talk about it, I say, "I am a corporate lawyer." I get the feeling they imagine me in some downtown office in a suit doing legal things they don't want to hear about.

It's true that I work in a downtown office in a suit, doing legal things you don't want to hear about—sometimes. But there is much more life in it than that.

So when people ask me what I do, *and I do want to talk about it*, I say, "I change things with words." This is as equally true as the first statement, but it always leads to a follow-up question: "What do you mean?"

Here is what I mean. When two companies have a deal they want to make—suppose one wants to buy the other—we turn the hope into a concrete reality with words. We move negotiations by choosing the most convincing words, we minimize risk by making sure a contract has all the right words, and at the closing of the deal we create a new reality by actually saying these words out loud: "Ladies and gentlemen, we are closed. Congratulations."

If you think about it, it's amazing. One moment there was no merger, and the next there is—simply because of words. Words create new realities. Small words have enormous impact.

As I look back on my career, it sometimes looks strange. I have at times been some combination of a missionary, a writer, and a lawyer. But when I think about words, it all makes sense. Mine has been a life of words. I've been in the business of words, trying to convince the world that there is truth, there is beauty, there is order. This has always been my vocation.

Understanding the power of small words is central to understanding the significance of daily prayer. We all desire to somehow shape our chaotic days into lives with meaning. *That begins with punctuating our days with words: the words of prayer.* I believe in the power of words—and especially words of prayer—to shape the world.

A WORLD MADE OF WORDS

The world began with words.

At first the majestic power of the voice of God spoke light into existence. Then came planets, plasmas, penguins, pineapples, and polar ice caps. When God speaks, the world takes shape.

Words bring order to chaos and form to the formless. But the power of words doesn't stop with God; their power passes to his image bearers—to us.

One of the most fascinating moments of the whole creation narrative is the moment when God handed the power of words to humans. God spent days moving in and out of a divine rhythm of creation and rest, speaking the world into existence and then sitting back to say, "Good! Very good!" And how did God make it all? By words.

Then God turns to humans and says, "Your turn," or, more formally, "Be fruitful and multiply." How did he pass on the task to us? By words. How are we to begin this task? Of course, by words. Adam's first job was to come alongside God and name the world. As the first poet—and zoologist—God invited Adam to work with him in speaking order into the world by the power of words.

So then, here we are at the beginning of a day. If we're to make something of the world, we must begin with words. Just as God framed the world in love, so we can use the words of prayer to frame each part of our day in love.

FRAMING THE MORNING IN LOVE OR LEGALISM

Every day of my life, I have woken with a kind of prayer. Depending on my stage of life, however, my prayers have been of radically different kinds.

In high school, it was, "Oh why, oh why, does first period start so early?" Or maybe "Please don't let anyone find out what happened last night."

In college, my prayers became inarticulate groans. Something like, "Oh please let it not matter that I'm missing that class."

In my recent life, these have made up my Book of Common Groans.

- I really should have gone to bed earlier.
- I really should have woken earlier.
- What the [insert choice expletive] are the boys doing up so early?
- Why am I always so tired?
- I seriously—*seriously* this time—have got to get that project done today.
- I shouldn't have watched that.

This is all to say that for my whole life, my day has begun with a profound sense of wishing something was different. Usually it revolves around what I've done or what I need to do. In this way, every day begins in a sort of prayer, no matter how inarticulate.

When I wake up thinking of what I've done, I often feel guilt over the day before. When I wake up thinking about what I need to do, I often feel anxiety over the day to come. Notice that in all these cases, my sense of things hangs on my performance. For most of my life, I've been framing the day with a sense of what you may call legalism.

Legalism is the belief that the world hangs on what I do and that God and people love me based on how I perform. This is an important concept because it's the exact opposite of the gospel: God loves us not because of *what we do*, but rather *in spite of what we do*—in spite of our good deeds and our bad deeds. Legalism takes the unmerited love of God and bends it into something earned—and just like that, the world is about us and not about him.

Legalism seems to be the default setting for human beings. So we will always wake up to some kind of prayer that makes the world about *us*—unless we create new habits of gospel prayers. These are prayers that make the world less about us and more about the love of God *for* us. Before I get to examples of these prayers, permit me a brief digression on prayer.

TWO KINDS OF PRAYER

The first kind of prayer names a reality that is. Like Adam's work of words in the Garden of Eden, it creates categories of meaning and names realities. This part of prayer is essential as it reminds us there are truths of the world: *God is good. We are loved. To be alive is beautiful. Gratitude is the way of happiness.* In this sense, prayer agrees with what God has created and reminds us of the way the world is.

The second kind of prayer is not simply naming what is but creating what can be. Just as God spoke mountains and soil into existence, so we use the words of prayer as a generative act of wishing new realities into existence. Often this work of prayer happens in the places where the order of creation has been broken: *Lord, have mercy! May your kingdom come. Please help!* Or, *Hold this widow as she grieves her husband's death.*

But not always. It may be *Lord, bless my children. May my neighbor know she is loved.* Or, *Guide my work today.*

The essence of all these prayers is to stand alongside God and speak order and love into a world that needs it. But our default morning prayers are usually broken versions of one of these two. We name false realities, or we create ones that shouldn't be.

ELECTRONIC PRAYERS

While I've been practicing some version of morning prayers my whole life, they radically changed when I got a smartphone. My smartphone exacerbates my tendency toward self-centered or legalistic morning prayers. Why? Because, of course, my phone is the portal through which the chaos of the world reaches my half-asleep heart through the pesky thing we call "notifications." This inevitably begins my day with all that I need to do and all that I've failed to do.

Our phones—and their programmers—are happy to set our habits for us. They would love to speak the first words of the day, and they usually do. Our phones—and whatever has come through them—thus shape the first desires of the morning and order our first prayers for us.

Before I banished notifications—emotionally prepare yourself now, for I will subsequently be recommending you do the same—I would wake to the prayers someone else wanted me to pray. If it was an early-morning work email with a task for me, I would begin the day wishing it could be done or that I could avoid it. If it was a news alert about some elected official doing something abominable, I would begin the day wishing people could just have some common sense like I did. If it was a social media alert, I would begin the day wishing my life happened in a tinted, square frame.

Each of these nudges invited prayers of their own, usually prayers that framed the day in stress, envy, or cynicism—and they are all the more powerful because they are done by unconscious habit.

KEYSTONE HABITS AND KNEELING PRAYERS

Habits are something we do over and over without thinking about them. They shape our world effortlessly. They form us more than we form them—and that's why they are so powerful.

A keystone habit is a super-habit. It's the first domino in the line; by changing one habit, we simultaneously change ten other habits.

Beginning the day in kneeling prayer is such a keystone habit. In morning prayer, we frame the first words of the day in God's love for us, which is to say we uproot the weeds of legalism that grow if we simply do nothing, and we lay the first piece of the day's trellis on which love can grow.

It took a terrible anxiety collapse to get me to think closely about what was happening in my heart each morning. I examined those moments, and I found that I was by default beginning the day by speaking the words of my pride or fear into each day. Framing the day in terms of me was effortless.

I wanted to change that, but changing habits of the mind is immensely tricky. Thoughts are slippery things. We can't grab them. Actually, we can't even touch them, and they often happen before we know it. Hence their power. They are the unnoticeable emotional water we swim in that ends up shaping everything.

For me, the first step in changing such thought habits was the discovery of the automatic "do not disturb" function on my phone. (Often reprogramming our phones is a way to reprogram our thought life, which goes to show how non-neutral our phones are.) Setting my phone to automatically go to "do not disturb" at 11 p.m. and stay set until 8 a.m. means that my family and other select contacts can call me if they need to. Anyone else who calls twice in an emergency will come through too. (Yes, that is sometimes work calls.) But otherwise it is silent.

This removed a thousand legalistic nudges during my morning, but alone it was not enough. That was only clearing the rubble.

The second step was to kneel. Often one of the only ways to take hold of the mind is to take hold of the body. As I kneeled, my sleepy mind was shocked into a new kind of moment. It wondered what was going on. *What are we doing down here on the cold floor?* (If you struggle with getting out of bed on time, this helps by hurting. Mild pain is a great way to stop snoozing. Fair warning.)

Now that the rubble is cleared out and the usual habits of the morning mind are disrupted, here we do what we were created to do. Like God, we speak words of love into the world.

Most days this is very, very short. When I wake exhausted or to a crying baby, often my first moment on my knees by the bed is very quick, a "Lord, have mercy." When I wake thinking about that big meeting or the project that needs way more hours than the day has, it's a slower pause for help: "Lord, I dreamt about this all night. I'm worried. Help me be like you and do good work that brings order to chaos." On those rare, glorious days when I wake rested, in disbelief that I got enough sleep and I have an hour before the kids wake up, it may be "Lord, I can't believe it! This is amazing! Thank you for your care for me! Let me love others today like you love me."

Here we recover the two uses of prayer, naming true realities: "God, thank you for another day I did not earn. You are so generous to me." And we are creating true realities: "Let me make something good of the world today. Let me love the world and all the people in it just like you love it."

In the smallest shift, the whole day has been reframed. Now it's time to get to work.

FRAMING OUR WORK IN LOVE

I have a kind of bipolar relationship with work. To this day I can't tell whether I'm fundamentally hedonistic or fundamentally a workaholic. Perhaps this is normal (but more likely it isn't). In any case, work is the place where I realize how much I am made to be like God and, at the very same time, how much I am not God. Because of this, meaningful work—even on the good days—is guaranteed to destabilize the day. This is why we need midday prayer to reframe the day in love.

Permit one last digression here. In order to understand why praying at work is important, we have to know what work is for.

To work is to be like God. This is because no matter what our profession is, work is where we make something of the world. To see why this is so Godlike, we have to return to the beginning again.

Go back to the opening act of the Bible. The spotlight has just come on, and we find the Trinity on the stage, working together to create this beautiful and bizarre material world. God is different parts blue-collar worker, artist, inventor, tinkerer, gardener, and entrepreneur, in all cases working with his hands, getting dirty, and calling this creative act good.

The Hebrew word that God says over and over is *tov*, which is something more than just *good*. It is something like *wow* or *whoa*—like the involuntary noise a stadium crowd makes when an athlete does something sudden and spectacular. Tov is the benediction God speaks over his creation, and if we miss that, we miss the fundamental truth that God is caught up in his love of matter and being and creation. In some way, we have to envision God like Jackson Pollock, the modern artist famous for throwing paint. God is slinging materials around, throwing paint at walls to see what sticks, and saying, "Wow! Tov! Tov! Tov!"

God's work is love. He loves the world into being. He sings it into existence, and he is enamored with the world he has made.

We can't understand what we're doing at work until we see that all of our professions are born out of the good work of God. Artists and inventors, like God, create things. That's tov. Lawyers and accountants name realities, bring order to chaos. That's tov. Builders make things that didn't used to exist, and plumbers fix things that are broken. They are both tov. Investors and entrepreneurs make things fruitful and multiply. It's all tov.

To work is to not be like God. Though all of our work is derivative of God's work; at the same time, we are so *unlike* God in our work. God speaks, and codes of DNA are ordered. God says the word, and magnetic poles tilt to their true north.

When we speak, it's much less majestic. Employees get offended, tasks get confused, and inboxes get clogged with useless new initiatives. We would like to say, "Let there be a third-quarter report!" But work is the place where we discover that anything worth doing seems almost impossible to do. Just today I was sending an email to someone requesting a deadline, knowing full well there was no way this person was going to honor my deadline. It's impossibly frustrating to know you're speaking meaningless words. Just ask a parent—it's exhausting and discouraging to waste words. We'd all rather have ultimate power.

In general, we aren't content to be like God; we want to *be* God. And this is the cause of so much self-centeredness in work. Work becomes about proving that we can accomplish something, that we are worth our salt, and that our voices are worth listening to—even if we have to bang the table or send a snarky told-you-so email.

So we invert the purpose of work. Instead of working as a way to love and serve others, we turn work into a way to be loved and served by others. Instead of longing to hear the "Tov!" of God, we work for the "Tov!" of people. And this is only the beginning of our brokenness; sometimes we actively labor to hurt people. Not only is the world complex and hard to manage, but evil abounds. Whether it's a competent bookkeeper working in the field of sex trafficking or an otherwise talented manager writing an email specifically intended to produce guilt and shame in an employee, often human work actively cultivates evil instead of love.

REFRAMING WORK IN LOVE

Kneeling prayer midday at work is a habit that reframes the workday in love because it resets the paradigm, often just when it's falling apart. I often begin the day in the office in a surge of energy and productivity. In the morning, tasks are clear. My to-do lists are organized. I have an (always unrealistic) sense that I'm going to accomplish everything that day. On the right cocktail of coffee, sheer willpower, and fear of failure, I can ride these plans for *at least* a couple of hours and sometimes even until noon.

But then the wheels begin to fall off. I usually notice this moment because I have the urge for still more coffee, even though I know it will give me the jitters. Or I have the urge to search the internet. For what? I don't know. I just want to search.

Somewhere between noon and two, I begin to realize that all the things I had hoped were going to get done are not going to get done. I realize that somebody is going to have to be disappointed. I begin to look down the barrel of the afternoon, and I see that I'm not good enough. I can't do it. People don't listen to me, and all the feelings of legalism return: *If I can't hack it, what am I worth anyway?*

Kneeling prayer midday is a chance to acknowledge that inexorable tendency and to reframe the day right as it is falling apart. At this point in the day, I close my office door and kneel. This is inevitably awkward. What if someone walks in? It's uncomfortable. Suit pants weren't made for touching knees to the floor. But these are good discomforts. They remind me something is happening. If I'm working in a public place, I may just set my hands on my lap and turn them up. I need something physical to mark the moment for my slippery mind.

My short prayers for midday often have to do with a confession that I've made my work about me. Then, hoping to rewire that impulse, I pray for a client or a coworker. Sometimes I try to remember the way our products or services affect people I'll never meet. Sometimes I think about how they affect people who can't afford them. Sometimes I pray for the unemployed.

Sometimes I just sit in silence and look out the window.

No matter what I do, the habit always interrupts things in the best of ways. By introducing a new habit, there's a hook in each day, a place where the focus on self is snagged and disrupted. And I'm reminded that work is not for me but for someone else, so I can turn the rest of my workday toward that someone, whether a client, customer, employee, or stranger.

FRAMING THE EVENING IN LOVE

Finally we come to the end of the day, the tenuous moment when we must take our hands off the wheel and let things rest.

One of my pastor friends told me he gets genuinely sad every evening because there's always so much more he wants to do with the day. I'm like this. I suspect we're all something like this. The evening brings me face to face with the reality of my limited life. There's so much we wanted to do or at least that we felt we should do. What seemed so simple in the morning seems so foolish in the moonlight.

We never could have done it all. We could work all night and not have the house clean, not have the project ready, not have the presentation prepared. And the baby is going to cry all night anyway, or your roommate is going to snore all night, or you're going to wake up at three, unable to go back to sleep. Pick your life stage.

We're frustrated because we had no time for free time. Or we're embarrassed because we squandered it all on free time.

The evening, then, can be a time of severe self-judgment. I often find myself lying in bed and facing the reality that I spent the whole day trying to justify my existence on earth. I lie there and find the scary reality hanging from the ceiling like a bulb that won't turn off. *Does any of it matter?*

That's a worrisome thought, and because of it, I want to tune everything out. And many of us do. A drink sounds nice; two sounds better. Sex sounds good; porn is easier. A conversation would help; but binging on TV will let me tune out. Catching up on reading would be restful; Twitter has some notifications that are probably more urgent. Lauren and I should spend some time talking; talking is hard, and there's a podcast of a sermon that everyone said we should listen to. Oh, and an article is trending. There are more or less healthy ways to escape, *but what I can't escape is the desire to escape.*

The exhaustion of a day places us into a twilight where it isn't easy to make the right decisions. We're tired, and because our bodies and our minds and our souls are all bound up together, we have trouble making choices. The business world calls it "decision fatigue." My dad sums it up in classic dad advice: "Avoid making important decisions after the sun goes down." The evening is a time of vulnerability. We haven't spent the day so much as the day has spent us. When our exhaustion gives way to our addictions, we're exposed for who we really are.

This is where an evening prayer can make one last and important turn in the day. Perhaps it's once the work is done, the dishes are clean, or the kids are asleep that we pause to pray, to purposefully frame the evening in rest instead of letting it slip into something else. Or perhaps we take seriously the act of going to bed and asking, *How am I going to end this thing?* Shall we lie awake in bed, letting all the replay tapes run? Shall we browse our phones for some recent celebrity scandal to bounce meaninglessly around our brain? Or shall we walk intentionally toward the rest we know we need?

No one can sleep while believing that she needs to keep the world spinning. But real rest comes when we thank God that we don't need to, because he does. Thus we kneel by the bed and place the period of God's mercy and care for us at the end of the day.

You made it through another day. It doesn't matter whether you feel spiritual or not. It doesn't matter whether you know what to say or not. It doesn't matter whether you've said the same thing every night for a month or not. It's just habit.

You say your prayers until your prayers say you. That's the goal.

BUILDING THE TRELLIS OF HABIT

When I was young, my mom planted Carolina jasmine in the garden next to our brick garage. Jasmine is a beautiful plant, but it is also a twining vine plant. If not directed, its prolific shoots spread toward all the other plants, overtaking and eventually killing them all.

My mom built a trellis next to the brick wall, directing the jasmine up and away from the other plants. After a few seasons, the yellow blossoms covered the whole wall. I still remember how that brick turned from something barren to something beautiful. I still remember the way the fragrance filled the backyard with the thick smell of spring.

Our lives are something like a jasmine plant, and our days and weeks are something like the trellis. At best, we're made to grow upward, blossom beautifully and fill the earth with all the rich fragrance of God's uncountable glories. Yet we are fallen. We are twisted. But that doesn't mean we don't grow; it means we grow sideways in ways we weren't meant to, often twisting into something that kills us and hurts those around us.

Should we do nothing, we will still grow. But we're likely to grow into habits that are destructive, not only to us but also to those around us.

Building the trellis of habit is a way to acknowledge the good ways God designed us as well as the ways the fall has broken us. It is a way to craft Annie Dillard's "net for catching days."[1] How else do we get our hands on time itself?

This begins with framing our days in love, and that begins with the words of prayer.

DAILY HABIT 1
KNEELING PRAYER AT MORNING, MIDDAY, AND BEDTIME

THE HABIT AT A GLANCE

The world is made of words. Even small, repeated words have power. Regular, carefully placed prayer is one of the keystone habits of spiritual formation, and is the beginning of building the trellis of habit. By framing our day in the words of prayer, we frame the day in love.

THREE WAYS TO START

Written prayers. You may want to begin by having a morning, midday, and evening prayer. Here are three you might use (these are available to print on the Common Rule website):

- *Morning.* Spirit, I was made for your presence. May this day be one I spend with you in all that I do. Amen.
- *Midday.* Jesus, I was made to join your work in the world. Please order the rest of my day in love for the people you have given me to serve. Amen.
- *Bedtime.* Father, I was made to rest in your love. May my body rest in sleep, and may my mind rest in your love. Amen.

Alarms and reminders. Once my friend Steve, after hearing a friend share for the fifth or sixth time how he wished he prayed with his wife, picked up the friend's phone and told Siri to set an alarm to remind him to pray with his wife. It was funny, but also common sense. Use alarms if you're having trouble beginning the rhythm. For a long time I had an alarm that went off at 1 p.m. each day at the office to remind me to stop and pray.

Praying with the body. Kneeling is a great way to mark the moment with physicality and humility. If kneeling is physically challenging or you are in public, try gently turning up your palms, setting them on your knees, or walking to a window.

THREE CONSIDERATIONS

Communal prayers. This habit often turns into a communal one. Some friends of mine who practice the Common Rule at their offices use this as a chance to take a break and pray briefly with a coworker. Usually they find an empty conference room and take a five-minute break together. My wife and I use evening prayer as a time to pray together before bed. Some mothers have told me they use the midday or evening prayer as a chance to pray with their children. Habitual, physical prayers are a great way to teach children the rhythms of constant prayer.

Variations. Because the habit of rhythmic prayer frames the day, consider adapting one point for a different time in your day. Morning commuters might place their morning prayer before starting the car or while driving. It's a good way to prepare your soul for the battle that traffic can cause. Consider praying just before going into your workplace or back into your home, as a way to transition your mind and heart.

Embracing repetition. Just because prayers are repetitive doesn't mean they're meaningless. Quite the opposite. Often these prayers form us over time because of their constant presence. Absolutely intermix spontaneous Spirit-led prayers into your day, but building the trellis of repetitive prayer is the way to encourage more prayers to grow.

SAY YOUR PRAYERS UNTIL YOUR PRAYERS SAY YOU.

READING AND RESOURCES

The Book of Common Prayer

Every Moment Holy, Douglas Kaine McKelvey

Common Prayer, Shane Claiborne, Jonathan Wilson-Hartgrove, and Enuma Okoro

DAILY HABIT 2

ONE MEAL WITH OTHERS

To the King!

A COMMON TOAST-PRAYER AMONG MY FRIENDS IN CHINA

So draw up your battle lines. Gather around this table, raise a toast to the King and the coming Kingdom, and fight back.

ANDREW PETERSON

Now, let us eat.

ROBERT FARRAR CAPON

FOOD AS FUEL

When I lived as a missionary in China, I had an American friend who was terrifically efficient at life. He was a voracious reader, hyper-rational, and very likable. The combination made him fun to talk to and hard to refute. For that reason, I was always interested in his theories on life and theology. I took all his book recommendations, and I even tried to adopt some of his life practices.

However, he had an Achilles' heel. I once heard him say that if he could take a pill each day for nourishment instead of having to eat, he would do it. I was scandalized.

The Chinese mastered the art of being "foodies" centuries before the advent of the everything-is-artisanal food revolution at the turn of the twenty-first century in America. *Everyone* in China is a foodie; there was no end of amazing places for Lauren and me to eat. I now realize that saying "Let's eat Chinese food" is like saying "Let's eat European food"—it doesn't mean anything because you haven't specified which region of China. Each region has a completely different kind of food—as different, for example, as French and Italian food—and every kind is amazing. What's more, we could eat like kings for under five dollars a day.

And yet, in this paradise of cuisine, if given the option, my friend *would choose not to eat*. This struck me as profoundly unnatural.

And yet the more time passes, the more I can relate.

EATING AS INCONVENIENCE

Anything worth doing sucks you in. If I have an important meeting or a big deadline, I wake up thinking about it, and all I want to do is get to the office to focus on it. Breakfast easily gets sacrificed on the altar of either my nervousness or my hunger for productivity.

Then, when I get into a surge of work, often the last thing I want to do is interrupt my flow to go have a snack. So there goes lunch. Then, of course, at about four in the afternoon, just when I would be ready to call it quits and go to a happy hour, the whole world realizes that there was something they wanted me to do that day. All the important requests come in, so getting home on time for dinner seems like a quaint tradition of a distant generation that didn't have email.

Even as I pursue good things like hard and focused work, I have a strange, unnatural wish to not need food. I power through a day on adrenaline and caffeine, thinking that maybe it's true after all that man does not live by bread alone. Maybe we can just live on a hunger for life.

Lauren thinks this is crazy. She wouldn't go more than a couple of hours without having a healthy snack by her side. And

I applaud her, because my way of doing it never lasts. Sooner (not later) it all collapses. The momentary productivity brought about by the breakfast on the go or the skipped lunch falls apart into an afternoon of a caffeine crash or a mood that can only be described as *hangry*. The result is rushing off to eat some fake food from a vending machine, which leaves me alone at a cafeteria table in the middle of the afternoon, feeling worse than when I began.

Many people plan their meals better than I do, but I set the bar pretty low. And I don't think I'm alone. In general, our culture puts busy schedules at the center of life and then tries to fit meals in around them. This is different from putting the table at the center and prioritizing our schedules around that.

In a culture of busyness, we seem to have made a strange flip. The solitary feeling of individual productivity and accomplishment is the necessity; time to stop to eat with others is a luxury. Of course we can't live without eating, so we make a concession to stop and stuff something in our mouths, as if food is simply a fuel—which is to say that our bodies are simply machines.

But we're not machines, we're human beings. A people who are made to eat. Regularly. And with others.

The daily habit of one meal a day with others is a way of moving the table back to the center of who we are and ordering our day around the kind of people we were created to be: *dependent and communal human beings*.

MADE TO EAT

The fact that we're made to eat says volumes about who we are and who God is. We *are not* just hungry bodies, nor machines that simply need fuel. We are hungry souls; we are people who crave the company and the delights of the table.

Our need for food says something profound about us. It says we need God, we need others, and we need the created world.

The need to eat reveals our dependence on God. Perhaps one of the most significant differences between us as image bearers of God and God himself is that we, unlike him, are dependent on things outside our selves. Food is our daily reminder of that. We were created to hunger because we were made to feast on God's generosity. Even if we attend to our appetite multiple times a day, it stubbornly comes back—almost as if it were strangely made to do so. This is not a hallmark of our fallen-ness, as if one day in the new heavens and the new earth we will get over this pesky need for food. Our hunger to feast on God's creation is actually a good way that we were made.

The need to eat reveals our dependence on each other. From planting to harvesting to preparing food, it's impossible to survive without each other's help. This distinguishes us from almost all other animals. Despite the fact that the modern world has obscured the way food comes to our table, every meal signifies an incomprehensibly vast web of dependence on our neighbors and their dependence on us.

The need to eat reveals our dependence on creation. We live in a web of mutual sacrifice. Whether you're eating plants or meat, every single bite signifies a moment when something died to give you life. We take that thing in, and it becomes our future life. There seems to be something distinctly Christlike about the fact that our ongoing daily life depends entirely on the sacrifice of other life on our behalf.

The table is the center of gravity. When we see food as fuel, we turn all of this on its head. We aren't grateful to God; we assume our right to food. We aren't grateful to each other; we create systems of food that embody the exploitation of our neighbors who grow, transport, prepare, and serve our food. We aren't grateful to creation; we consume the earth's food greedily and carelessly, as if the world were ours to binge on and trash instead of ours to steward and cultivate into flourishing.

Disrupting that fallen culture by recentering life around the table is not the responsibility of a few health and farming activists. It is the call of every human made in the image of God. The Christian call to love neighbor should not and must not ignore the daily ways we eat, what our food is, or how it gets to us.

Given that our communal life revolves around our need to eat, we may say that the table is the center of gravity for loving neighbor. The daily habit of eating at least one meal with others is important precisely because it asks us to rearrange our priorities around the communal table and to acknowledge that we are made for food and for each other.

ORDERING SCHEDULES AROUND THE TABLE

When Lauren and I were leaving China, we had almost no money. Our plan was that she would continue her career in philanthropy consulting and I would start law school. (In other words, she would make money while I spent it.) So when we moved to DC, one of the most expensive housing markets in the United States, our wallet was saved by a dear friend who invited us to live temporarily in a communal house called the Brethren House.

The way of the house was simple. You paid very little in rent, but you had to cook and clean and come to meals. There was a meal every evening, and you were to be there. Period. Once a week you were to make the meal, and once a week you were to clean up after it.

As you may expect, this was incredibly countercultural in the chaotic world of DC power-schedules and appointments. I found myself jumping on the metro to head home just as all my classmates were getting started on the next round of studying or meetings. But I had to pay the rent by coming to the table.

This was profoundly formational. In a matter of weeks, we became remarkably close to these people, *not* because we spent so much time around them but because we came to the table rhythmically. Because of the centrality of the table in our daily schedule, our lives were calibrated for relationship instead of for loneliness and busyness.

Scheduling food. Think about how difficult it is to organize your life around a daily, common meal. You must sync schedules, plan, buy food together, have mutual understandings about food values and tastes, and refuse an extra hour of work or other productivity to make it to the table on time.

The fact is, we have to rearrange our lives to have a meal a day with others. This is not an awkward reality to brush aside while talking idealistically about new habits of eating; it is *the point.* Our schedules *need* to be bent around the common table. The daily habit of a meal with others forms us in that direction.

Even if we fail constantly, which we are bound to do, the very act of trying to have one meal a day with others sets us firmly in a countercultural current. But this current actually pushes us *toward* our neighbor while the usual cultural currents push us to the fake meals on the go and all the health problems and loneliness that come with them.

Working through meals. When I was a new lawyer in a big law firm, I used to be petrified of saying no to anything asked of me. As a result, I never *ever* knew when I was coming home. I would check in with Lauren at five in the evening. Sometimes the news was "I'm on the way." Sometimes it was "I'll see you in the morning." We had two baby boys at that time, and my lack of schedule was a continual point of madness for our family life. No one, including me, knew what to expect.

Some time after my anxiety crash, as my wife and friends helped me craft a schedule designed for peace instead of madness, Lauren and I found that setting a generally consistent family dinner time was a way to place an anchor of community in the day. It signified the end of work, even on days I wasn't ready to end work.

We began to cultivate this daily rhythm of one meal together, and not just for the sake of the family. It was as much for the sake of the sanity of our own schedules and to honor the limits of what was actually possible in a day.

Like most limits, we quickly realized it brought more freedom than restrictions. I knew that *sometimes* work would be crazy—that a merger would be closing, so I would have to work around the clock. That was fine. That's what I got paid for.

But like all work on formational habits, the background norms are far more powerful than the exceptions. The pattern, not the anomaly, is the key. The new norm was that I began to tell people that I was going home around six for family dinner, and if they needed me later, I'd come back after bedtime.

This was a game changer. Suddenly not only did my family know what to expect, but my coworkers and clients did too. No one ever said anything bad about it. They still haven't.

The schedule now revolves around the table, not the table around the schedule.

ORDERING SPACE AROUND THE TABLE

When my wife and I bought our first house in Richmond—the house where we still live—she tried to convince me we should buy the perfect, long, vintage teakwood dining table. It had popped up on her Instagram feed, put on sale by a local art-deco dealer who was selling everything and closing her business. It was the kind of table that takes up a lot of space and has a grain that calls attention to itself. It begs to be the center of a room.

At first I balked. I tried to point out gently that this table—despite being on sale and actually a very good price—would nonetheless cost over 10 percent of our shoestring renovation budget. But she was adamant, pointing out that it was well worth the expense, because it was the kind of table that a family gathers around. It was the kind of table you host neighbors at. It was the kind of table you create memories around.

Of course, we bought it. And of course, she was right.

The table quickly became the centerpiece to build the rest of our home around. Our family of six eats there daily, but we can insert leaves to extend it to fit sixteen or more people for gatherings.

I didn't have any words for it at the time, but I now see that what Lauren did was argue for a center of gravity for our household space, which is worth way more than 10 percent of any renovation budget—because the table is where life happens. It's where a household learns to love.

THE SCHOOL OF LOVE

Regardless of whether you live in a house with family, with friends, or alone, your household is not just a place where you spend a lot of your time. It's a place of formation. For that reason, generations of Christians have described the household as "a school of love." The point of that phrase is to emphasize that most of the intangible things that make life worth living are learned at home. Or not.

The household is where we're first introduced to what it means that the fundamental goal of life is to love each other. The table is the centerpiece of this formation.

Think through all the ways the values of love are communicated over food. We serve each other. We clean up after each other. We take turns. We share. We fight and forgive. We praise and compliment. We express gratitude. We tell stories and ask questions. We listen. We hear each other pray.

If our household routine is too busy to allow meals that unfold with family members, roommates, or neighbors, we should at least admit that the school of our household is not actually a school of love but rather a school of busyness. That house is a place where we teach each other how to do too much and be stressed about it. Or if the kitchen is way too messy to produce food, and the table is way too cluttered to sit and eat at, it's unsurprising that we end up being formed in loneliness, eating by the TV or the computer. The norms of our table signal the norms of our community.

How we order our space affects how we order our relationships. And just as we order our schedule around the table, ordering household space around an inviting table takes an immense amount of work.

What astounds me about being a parent is the amount of time my wife and I spend clearing laundry and kids' stuff off the table, preparing food, serving food, cleaning up after food, and beginning again to prep food for the next day. It is *totally* endless.

Many years ago, I recall someone during a mission trip telling the people who set up chairs, "Don't you belittle what you're doing. You are creating viewpoints for the gospel." As trite as it sounded to me at the time, I never forgot that, because I couldn't deny its truth. The table is similar. An endless cycle of chores is required to create a place for food and conversation, but that seat is where love is. What's more worth our time?

I now see this overwhelmingly repetitive and unending chore list of cooking and cleaning, wiping and sweeping, as a Christlike activity of creation and service. We create with liveliness, beauty, and flavor. And then we order. We turn the chaos of dishes into clean counters and the smatter of leftovers into neatly boxed lunch containers. We get breakfast ready, we share a drink, and then we turn out lights to get ready to do it again and again and again. And so we rest. It's for love, and it is tov.

This is the only way to view the endless tasks of the table as a calling, not as a waste of time better accomplished by a microwave or a drive-through. Rather, it's stewardship of the tools of the household, placing them into service in the school of love. This is incredibly difficult on any one evening but incredibly rewarding as a habitual way of life.

Yet we have only just begun to love. The amazing thing about making the table the center of gravity is the way we suddenly find that it begins to pull others into its orbit.

DRAWING THE NEIGHBOR IN

I grew up in a family of six children, all of whom now live in Richmond. As of right now, there are four spouses, a dozen grandchildren, and two dogs. While it's amazing to have my parents and all of my siblings in the same city, we quickly realized that the reality of our lives and schedules would make it hard to coordinate a gathering where everyone was free—unless it was a regular time we could all prioritize and plan on.

So a couple of years ago, we began a family tradition of gathering all twenty-four of us (and still counting, but not counting the dogs) for a Sunday family lunch. This is a way for us to find a spot in the week where we can come together in an expected rhythm. All the cousins can play, and all the adults can eat and talk.

While this habit of Sunday family lunches has been a good rhythm to pull us all in together, it is also potentially exclusive.

Sunday afternoons is often a time my neighbors are free, a time when people I meet at church or work can get together. Saying that I *always* eat lunch with my family on Sundays can be a way to wall myself off from someone who is lonely, who doesn't have family or friends around.

In the past year or so, it occurred to us that when you're already setting a table for twenty-some people, setting out another two or three plates isn't a big task. So we've started inviting others every couple of weeks. Each of our past three weeks have seen multiple guests at the family table.

And amazingly they blend right in. We've had guests ranging from a neighborhood kid from the projects down the street to an elderly couple who just moved to town. We've hosted old friends who know our family. The other day a new Richmond resident whom I'd just met joined us.

Opening the household table on a regular basis creates an undercurrent of the Christian life that mimics the adoption ethic. The family is open, not closed. And there are few more precious gifts to give someone who is lonely than inviting them into the circle. This doesn't happen automatically; but on the other hand, once you have the rhythm of the table, all it takes is an extra chair. When we combine habits of communal rhythms around the table with habits of maintaining open seats there, we come to one of the most sweet and powerful ways of living the outward life, a life inclined toward the love of our neighbor.

Cultivating the rhythms of the table *must not* pull us away from the outsider. Fortunately, it does not have to. A redeemed table is one that invites outsiders in. Perhaps even more importantly, it invites them into a place worth joining. If we do not intentionally form communities of eating, there's not much to invite anyone to. But when we cultivate rich community over food, it says something more powerful about God and God's people than could be put into words. As Christine Pohl wrote, "How we live together may be the greatest sermon we preach."[1]

HABITS AS LIGHT IN A SECULAR AGE

One of my favorite cultural critics, Ken Myers, argues that the kind of atheism we experience in America today is not a conclusion but a mood.[2] This is an incredibly important observation. If secularism is not a conclusion but a mood, we cannot disrupt it with an argument. *We must disrupt it with a presence*.

The truth is that we live in a culture where most people are remarkably resistant to hearing verbal proclamations of the gospel. What's more, it seems some of them really *can't* hear it. We no longer share a common vocabulary for communicating whether truth exists, what can be called good, and what love means. But that is okay. God is not alarmed. Our secular age is not a barrier to evangelism; it is simply the place of evangelism.[3]

Ever since returning from China, I've had an abiding interest in asking this question: "How is it that the West can be re-evangelized?"[4] One of the reasons I'm so compelled by the life of habit is that I see habits as a way of light in an age of darkness. Cultivating a life of transcendent habits means that our ordinary ways of living should stand out in our culture, dancing like candles on a dark mantle. As Madeleine L'Engle once wrote, "We draw people to Christ not by loudly discrediting what they believe . . . but by showing them a light that is so lovely that they want with all their hearts to know the source of it."[5]

I will try to show how each of the Common Rule habits for the love of neighbor act as such a "light so lovely," and that begins at the table.

In a secular age, eating may be our best chance for evangelism. Rosaria Butterfield described this powerful ethic in her book *The*

Gospel Comes with a House Key. She herself was converted from being an aggressive antagonist of the Christian faith to a devoted follower of Jesus through friendship with a pastor and his family who continually invited her to their table.

Don Everts and Doug Schaupp write in their book *I Once Was Lost* that one of the main things our neighbors who don't know Jesus need is simply to trust a Christian.[6] That begins at the table—a table lovingly set with good conversation and an extra chair.

A CULTURE OF COMMUNION

More Americans regularly eat alone now than ever before. Food is meant to bind us to God, neighbor, and creation, but we live in a culture where our eating habits keep us apart and increase our isolation. The best way to understand the Common Rule habit of one meal a day with others is to see it as a way of turning on that light of presence in a dark culture of loneliness.

Like all of the habits, the point is to adopt the rhythms of the gospel into our daily lives and to have those rhythms become a blessing to us and our neighbors.

The central promise of salvation is that because of the death and resurrection of Jesus, God and people will eat again. The end of the world culminates not in clouds and harps but in a feast. At the wedding supper of the Lamb, the divine presence is restored to us over a table of food.

But we don't get invited to the table because of anything we've done. We get invited because of what Jesus has done. This is why Christians regularly come to the Communion table to feast on the body and blood of Christ. It is a reminder that because of Christ, we will commune with God again over food.

To find each other over food each day is to plant and cultivate the culture of communion in our households, offices, and neighborhoods. My friend Drew likes to say that the work of the kingdom begins at the table. I agree. So let us eat!

DAILY HABIT 2
ONE MEAL WITH OTHERS

THE HABIT AT A GLANCE

We were made to eat, so the table must be our center of gravity. The habit of making time for one communal meal each day forces us to reorient our schedules and our space around food and each other. The more the table becomes our center of gravity, the more it draws our neighbors into gospel community.

THREE WAYS TO START

Family meals. Getting into the rhythm of a family breakfast or dinner may be the best way to start this. Pick which meal works best for your household, and try to make it the anchor of your schedule.

Standing coworker lunch. Establishing a regular lunchtime hour with coworkers can be a great way to create a meaningful break in the workday as well as to build relationships. Try getting a group together and picking a time. Everyone doesn't have to make it every time, but having a standing hour where others know they can eat together goes a long way toward building a habit and toward creating a culture of community. Sometimes it's easier to invite a new or lonely coworker to a standing group than a one-on-one.

Eat communally while alone. If you wish you had family or friends to eat with but find yourself alone, this could be a hard chapter for you to read. I want you to know that the supper of the Lamb awaits you, and there will be a day all loneliness is gone forever. One of my single friends makes it a habit to eat at the same restaurant counter at least once a week. He's a regular there and always talks to the servers or others that sit down at the bar. While you may be in pain, consider trying to be a blessing to others by having a regular rhythm of eating somewhere

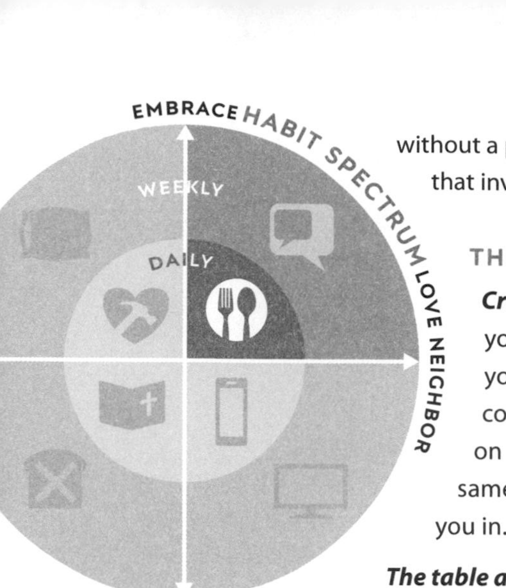

without a phone or headphones—in a way that invites conversation.

THREE CONSIDERATIONS

Creating space. Space matters. If you're at home, get the detritus off your table, which is for food and conversation. Maybe put a candle on it. If it's a bar or island, do the same. You want a space that draws you in.

The table as formation. Some of the ways my family tries to make mealtimes intentionally relational is by lighting a candle, having prayers everyone knows so we can take turns, and having regular questions to ask and answer. In the morning, ours are "What are you hoping for today?" and "What are you not looking forward to today?" At dinner, it's "One good thing, one bad thing, one funny thing." With friends, I've always appreciated the One Conversation Rule. That means, at some point in the meal, everyone has a single conversation instead of lots of side conversations.

Turning meals inside out. My friend Tom tries to hold Friday nights open. His family eats the same inexpensive and simple homemade pizza meal. That way it is low prep, and no one has to think hard. Each Friday, they invite someone new. This is a great way to take the family rhythm of dinner and flip it open to a neighbor. Another way is to eat in your front yard or on your front porch instead of your backyard or back porch.

> SECULARISM IS NOT A CONCLUSION; IT'S A MOOD. THAT MEANS WE CAN'T DISRUPT IT WITH AN ARGUMENT. WE MUST DISRUPT IT WITH A PRESENCE. THIS IS THE ESSENCE OF THE TABLE.

READING AND RESOURCES

The Supper of the Lamb, Robert Farrar Capon
The Gospel Comes with a House Key, Rosaria Butterfield

DAILY HABIT 3

ONE HOUR WITH PHONE OFF

"I will be hidden from your presence,
I will be a restless wanderer."

CAIN

Play with me.

MY CHILDREN

"Behold, the dwelling place of God is with man."

A LOUD VOICE FROM THE THRONE (REVELATION 21:3)

Justin?

MY WIFE

I AM NOT HERE

It's two in the afternoon when the office line rings. Lauren is calling for our usual afternoon check-in. I imagine she's just gotten the boys down for a nap or perhaps just left them with a sitter and is on the way to an afternoon meeting. I close my laptop and put her on speaker phone.

As we talk, I walk around the office. She's updating me on her morning.

The Washington Nationals eked out a win last night, but barely. And of course the president said something inflammatory, which all the media stations—instead of ignoring—have decided is a national crisis.

Dan has texted me about the article that we discussed last night, and I remember that I need to get back to Chris about whether we can do drinks next Thursday. But someone retweeted me, so maybe comment on that first or—

"So what do you think?" Lauren says, snapping me out of it. I have zero idea what she had been saying, because I had subconsciously picked up my iPhone and begun scrolling notifications.

It is hard to tell which is more embarrassing for me, that I ignored the love of my life or that this happens all the time. When I stop to think about this habit—whether it was a call with my wife or a conference call with a client—I always feel vaguely guilty. I tell someone that I'm talking to them, but then I give my attention elsewhere. Perhaps I've lied to them about where my presence is, or perhaps I'm lying to myself about how many ways I split my presence and can still be present.

Either way, I feel the discomfort of a small deceit. I've tried to be two places at once, and as a result, I was no place.

This is the core struggle of the smartphone. It's amazing because it allows us to communicate our presence across time and space, but it's dangerous for the very same reason. It can fracture our presence across time and space until nothing is left. Usually this happens simply by habit, like me talking via phone to my wife while doing two or three other things.

We don't mean to live lives of absence, but without meaningful habits of resistance, smartphones are impossible *not* to look at. If we do nothing, we're sure to live a life of fractured presence. And that's not much of a life at all, because presence is the essence of life itself.

PRESENCE AS LIFE AND VICE VERSA

Presence is at the heart of who we are, because presence is at the core of our relationship with God. From creation to salvation, the story of the Bible is fundamentally a story of presence. Eden was Eden because the unmediated presence of God was there. God was *with* Adam and Eve, until sin broke the bliss of that presence.

After eating the fruit, Adam and Eve suddenly wanted to cover themselves with clothes and hide. This is the hallmark of life as we know it now. We hide from each other, and we hide from God. We long for the face of God, but we can't bear his gaze either.

Sin has turned a people meant for presence into a people of absence, but fortunately the story of the Bible doesn't end there. Since humans first walked east of Eden, away from the presence of God, God has spent the rest of the story tracking his people down.

First he followed them through deserts and wildernesses. He appeared in clouds of smoke and burning bushes, he found them in midnight dreams and pillars of fire. He manifested his presence on a mountain, in a tabernacle, and in a temple. The Israelites are known as God's people because of one thing: *They had God's presence among them*.

This culminated into the salvation story of the New Testament. Jesus is called Emmanuel because it means "God with us." Jesus came to see to it that God and man could be together again. He did this through his death and resurrection. By atoning for our sins on the cross and breaking the back of death in the resurrection, he cleared the path for the presence of God to once again become the cornerstone of our reality. Now, like the Israelites, a Christian is defined by "God with us," and this is why we have the Holy Spirit.

What's more, our great hope is to consummate this presence. In the kingdom to come, God will look at us, and we will look back. In his gaze, we will find the definition of our own lives and indeed the definition of all things.

That is why, for a Christian, presence is the heart of everything.

THE SMARTPHONE AND FRACTURED PRESENCE

One of my favorite concepts in the Harry Potter books by J. K. Rowling is Voldemort's horcruxes. In trying to conquer death itself, Voldemort splits himself into seven different horcruxes so if one is killed, he still has life elsewhere. What he doesn't realize is, that in trying to multiply his presence everywhere, he splits his own soul. In the end, the very effort to be omnipresent is the cause of his absence.

Our efforts to split our presence are similar. As image bearers of God, we have a powerful presence to give to others. But unlike our omnipresent God, we have a limited presence. To long to be *omnipresent* is a false, bent longing to be God himself. It is not the way we were made to be. And like all efforts to be God, it will break us.

When we try to be present everywhere, we end up being present nowhere. When we try to free ourselves from the limitations of our presence, we always become enslaved to absence. But when we embrace our reality of being able to be present only in one place, we find the deep joy of being present someplace.

This is why we must be attentive to our smartphone habits. The smartphone is a tool that enables many things, but it will never multiply our presence. When we try to use it that way, it only brings absence, and this absence is the cause of much brokenness in the world.

Think of all the ways we now use our smartphones to fracture presence: working while vacationing, checking emails on a date, sexting with someone we'll never meet, taking calls while playing with our kids, interrupting our dinner with news notifications, posting a conflict instead of talking to someone about it, taking pictures of people in distress instead of helping them, taking a picture of someone who doesn't know it, watching videos of someone who doesn't want to be watched, curating our whole lives on a media feed in order to be "with" everyone except the ones we are actually next to. These are all ways of fractured presence, and they do real harm, both to us and our neighbors.

HABITS OF RESISTANCE AS ACTS OF PRESENCE

The goal of resisting absence is to give others our presence.

Turning your phone off with family and friends. My brother-in-law, Dan, told me about an evening at home with his two sons. They were having a living room sing-along to the soundtrack of *The Lion King* (which, incidentally happens to be my idea of a good Tuesday night).

Unfortunately, right in the middle of their crooning, Siri came on Dan's phone via the Bluetooth speaker, announcing that she was about to make a call to someone random. Somehow their permutations of the "Circle of Life" lyrics (which, honestly, who knows what they are saying) sounded just enough like "Hey, Siri" to call his phone to attention, which was sitting in the kitchen.

Little James and Eli stopped dancing while Dan dashed in to keep his phone from calling some decade-old contact in his address book and giving them a private audience to the evening's revelries.

When Dan told me this, we laughed about it. It wasn't so much that the phone ruined their moment of play—they got right back to it—but when we thought about it, the interruption was as interesting as it was humorous. For Dan, it was a moment of epiphany. "Our phones cry out for our attention," he told me. In this case, literally.

Attention is our precious commodity. Our life is defined by what we pay attention to. This means our life is defined by which of the many cries for our attention we heed. If we're going to take that call seriously, we have to acknowledge that our phones are carefully designed to attract our attention and sell it to advertisers.

There is a powerful monetary incentive that frames the functionality of all our devices. It doesn't necessarily make them evil capitalistic machines, but it certainly means they aren't neutral in the slightest. And that means we have to do the hard work of governing them, because they will not govern themselves, and they would love to govern us.

Studies show that having your phone on silent or in the room is far different from having it off or out of sight.[1] This is why the Common Rule habit suggests turning your phone off for an hour.

Try it right now. Just as an experiment. I dare you. Wait until it is off to read the next paragraph.

First of all, you almost certainly just got distracted by a few notifications on the way to turning it off. If you're back, I'm impressed if it took you less than five minutes.

Second, now that your phone is off, you probably notice a palpable feeling of aloneness—like someone walked out of a room. It very well may include a tinge of anxiety or dread, because if we turn our phones off, that means we cut off the possibility of our presence from others. We can't reach or be reached. This is exactly what is scary, and it's exactly why we should be turning our phones off every day as a habit.

The goal is to regularly cut off the ability to be reached by everyone and anyone, so that in those limits we can be fully present to someone. For me, this is often my kids. Playing with children takes so much concentration and energy, and they always, always know when you're multitasking. In fact, I've noticed that my son, Coulter, who is great at getting my attention by use of his supersonic squeal, doesn't use said squeal when I'm looking at my phone. Instead he hits something, sometimes me. He never does that any other time. He can't talk yet, but something in him already knows that my attention to a screen is different from my attention to a frying pan or a guitar, and he responds accordingly.

My hour with my phone off starts shortly after I get home from work. This is one of the hardest times of the day because my children are ramped up for my attention, but I'm still trying to come down from the workday. My habit is to get changed, make one final email check to make sure things are in order at the office—and if not, to tell someone that I'll get back to them later that night—and then to turn it off and put it in my dresser drawer. It's a weird feeling, almost like hiding a valuable under a mattress. You walk away but your mind stays on it. You can visualize it sitting there in the dark.

But whether the boys and I are riding bikes to the park, initiating a royal rumble on the living room floor, or setting the table together, my presence is fundamentally different that hour of the day. I am with them. Whatever we're doing, it is together.

Being a parent is so very hard because of the infinite demands on your attention. Yet my deep desire is to rise to the occasion. At my best, I want to be a reflection of their Father in heaven. When they look up and squeal, "Play with me!" I want them to find my gaze there, looking back down at them—not buried in a phone so they feel they need to earn or interrupt my attention by acting out. I want them to know they have it. So when I'm with them, I am actually *with* them. They have my gaze, which is to say they have my attention, which is to say they have my love.

This is no different from what I long to give friends and acquaintances. When we come to the table together, for example, we come together to give our attention to each other. It's a perfect time for the hour with no phone access.

As you may expect, I do use my phone around my children and friends all the time. There are days when I make a judgment call to keep it on or beside me at the dinner table because something at work is urgent and the phone is the only reason I'm able to leave the office. I also don't turn it off when I'm away from my wife (unless she knows about it).

If I have to use it while talking to someone else, I try to make it a rule to excuse myself and tell them what I'm doing. There's a huge difference in presence between mumbling an "mm-hmm" and checking my phone while someone is talking, and saying, "Excuse me, I need to let my wife know what time I'll be back."

Lauren and I do this with our kids too. There are plenty of times when there's a really good reason to ask them to wait while I do something on my phone. Our habit is to tell them what we're doing and why it's important that they wait until we've finished. "Are you finished with that important work email yet?" is a question I hear from my sons often. But it isn't because I'm always on my phone; it's only because when I am, I tell them exactly why.

If I'm not willing to tell someone why I'm asking them to wait, usually it's because there isn't a good reason to ask them to.

Turning your phone off at work. Once a client called me on a Friday afternoon and asked me how my schedule looked on Monday. I breathed a sigh of relief. Whenever someone asks me about my schedule, I get nervous, so I was just glad he hadn't asked me how my schedule looked "this weekend."

His company was acquiring a Canadian engineering business, and he wasn't entirely confident that the Canadian law firm who wrote the agreement had structured it properly. We had done a number of domestic acquisitions for him, so he wanted me to double-check their work. "They'll send me the agreement on Monday morning," he said. "If I send it right to you, how long would it take you to review and mark-up? It's ninety-plus pages."

Important interjection: In the past, a ninety-page stock purchase agreement has taken me upwards of three days and a significant coffee budget to get through.

"Three or four hours," I said.

"Great," he replied, "because we need your comments by early afternoon Monday."

On Monday morning, when the agreement arrived, I printed it, shut my computer, left my phone on my desk, and walked two blocks to a coffee shop with only one pen and the stack of paper. I felt all the usual anxieties of being unavailable, but I told the client that I would do good work for him. I knew that to follow through on that promise meant not only saying no to other requests that morning but also protecting my brain from distractions.

Within an hour, I felt the intense pleasure that can come only from doing focused work on one important task—often referred to now as deep work or flow.[2] This is the state where real work happens, and it never happens in the presence of a phone. I knew people were calling my phone and new emails were landing in my inbox, but there was the afternoon for that. And I've learned that when I don't silence other options and focus on one thing, that one thing can morph from a three-hour focused task to a three-day sprawling project littered with needless distractions and interruptions.

At noon I walked back to the office with an "issues list" handwritten on the back page. I typed it up and emailed it, and then began returning phone calls.

I tell you this story not because it is typical of me but because it was an unusual victory. So often I've had the same request asked of me, but it ended differently. By keeping my phone with me, I've gotten distracted by urgent requests masquerading as important requests and have ended up coming back to a client with nothing but an apology and a sheepish ask for more time.

We send these kinds of emails to each other every day: “Sorry—crazy morning, something came up. Can I get you that report tomorrow?” Every once in a while, this is actually a true statement. Almost always, however, it’s a cover-up. It wasn’t the morning that was crazy; it was me that was crazy. I made it crazy due to my inability to be present to work with focused attention.

In the age of smartphones, the ability to resist distraction purposefully is not just becoming the single most important career skill, it’s also a matter of whether or not we love our neighbors through our work. If it’s true that our work is fundamentally good because, like God, we order the world to serve our neighbor, then the question of neighbor love is this: Am I too distracted to actually serve my neighbor?

Love can be expressed in all sorts of ways. Listening to a coworker at the water cooler. Being attentive to whether your human resource policies treat people more like resources than humans. Learning to give encouragement and constructive criticism instead of just tearing someone’s work apart.

These are all wonderful, but there’s something else even more fundamental: Are we actually doing good work for our clients, our customers, our supervisors? Are we giving them the one thing that produces a great product or service: our undivided attention?

There’s no love of neighbor outside of attention to neighbor. This is true on our cul-de-sacs and in our offices. Having periods of keeping your phone off at work is the keystone habit for loving neighbors through your work.

Turning your phone off to seek silence. I was a couple of months into my anxiety crash when I picked up a newsletter from my friend Josh that had come in the mail. As I looked over the pictures, I saw Blaise Pascal’s famous line in a box quote: “All of man’s problems stem from his inability to sit quietly in a room alone.”

I immediately turned the letter facedown on the kitchen island and walked away as I felt a rush of panic come over me. It was as if the words had hurt me, and I needed to get them out of sight. Why was I afraid? Because I knew it to be devastatingly true.

The worst depths of my emotional breakdown happened when I began to fear my own mind. In general, whenever I'm alone, the noise of my own mind blares like a Niagara Falls of random desires and self-condemnations. The thoughts seem uncontrollable, and during the periods when I suffered panic attacks, the volume of that noise increased to an unbearable roar.

The tragic irony when it comes to our troubled emotional lives is that distraction functions both as one of the best quick fixes *and* as one of the roots of the problem. When the distractions fade away and the roar of silence begins, we're confronted with the question that haunts us: Who are we really, now that no one is looking?

To sit peacefully in silence requires knowing your soul, knowing who you really are, and being fundamentally okay with that and at peace with that. This is exactly why we avoid it; we don't know who we really are. Or if we do, we're terrified of ourselves. Silence confronts us with that fact, so we will do anything to avoid it.

In the couple of years after reading the Pascal quote about being unable to be in the quiet, I began to realize that I needed to confront my fear of silence. I needed to not be afraid of boredom. I needed to cultivate the strength to stare at a wall without my heart rate going up.

Around that time, I started going to a local monastery-like retreat center called Richmond Hill a couple of times a year to join others in a day of silence. The first time I walked into the room and the retreat leader told us that we were going to begin with twenty minutes of group silence, I was genuinely afraid. The beginning was horrible—it was as if my mind was storming the beaches of Normandy. I was shredded to ribbons with the bullets of self-loathing and self-doubt.

And yet I survived.

Anything that is good for you initially hurts. Silence is the same. But by the end of the day, it seemed as if the silence itself had begun to loosen the tangled knots of my heart. Now I see that that was exactly what had happened, because constant distraction and inattention is what had tied me up in knots in the first place.

SILENCE AS NEIGHBOR LOVE

Silence begins as a personal practice, but it always ends as a public virtue. Just think of social media. It exists in the form we know it because we don't know who we are before coming to it. When we can't answer the question of who we are in silence, we can't answer it in public either, and our insecurities spill out into the world in the form of manipulations. We hide our confusion behind a posture of perpetual offense. If we are opposed to someone or something, that's enough to create our identity for the day, which is to say we use others so that we can get the temporary identity we need. We don't know who we are, so we make others feel the pain of our insecurity.

Only when we know who we are can we turn to love others, not use others. Only then can we actually listen to them. As Kyle David Bennett writes in his book on how the spiritual disciplines are for the love of the world, "How can we love our neighbor if we never allow her to reveal herself because we are always chattering?"[3]

Even more, when we cultivate inner rhythms of silence, we become attentive to the voice of conscience, to the voice of God's love for the world, and to the voice of our neighbor's need. But it's possible to avoid confrontations with our conscience simply because we're never quiet.

Imagine Martin Luther King Jr. in his kitchen in the middle of that night. He couldn't sleep because of the death threats and other pressure. What if he was so distracted by his smartphone, which he would have understandably pulled out to numb the pain, that he didn't hear the voice of God saying, "Stand up for justice, stand up for truth; and God will be at your side forever." What love for neighbor we would have missed.

Who knows who God is calling to be the next prophets of race relations in our country? I imagine they're more attentive to his quiet voice than to the noise of social media. I pray they will be.

CULTIVATING A LIFE OF PRESENCE

When I think about my daily afternoon check-in with Lauren, I feel the wonderful paradox of marriage. Marriage is remarkable because of the ways the boring, common, daily routines tie two lives together in a way that they can never be separated. The common gives birth to the uncommon. The ordinary paves the way to the extraordinary. With no one else do I touch base every day at two in the afternoon to mumble about how the day has been, kick around evening plans, and end with "I love you too."

This beautiful, nearly unconscious routine would never be possible without a phone. And yet often my unconscious interaction with my phone threatens the life of it.

Use your phone one way, and it fuels the life of love and presence you long for. Use your phone the other way, and it robs you of everything you were made for. But remember that the phone isn't neutral. We can't use it the right way without habits that protect us from the wrong way. When we do nothing, they tilt us toward absence.

This is why we must cultivate habits that resist absence—because we were made for presence. Cultivating the daily habit of turning your phone off for an hour each day is the keystone habit that can change the way you think about your phone and spark new daily routines that usher in a life of presence.

DAILY HABIT 3
ONE HOUR WITH PHONE OFF

THE HABIT AT A GLANCE

We were made for presence, but so often our phones are the cause of our absence. To be two places at a time is to be no place at all. Turning off our phone for an hour a day is a way to turn our gaze up to each other, whether that be children, coworkers, friends, or neighbors. Our habits of attention are habits of love. To resist absence is to love neighbor.

THREE WAYS TO START

Hour at home. I find that having the same hour every day goes a long way toward creating a rhythm of presence at home. In my house, somewhere around 6:30 and 7:30 p.m. is the best time for phones to be off for conversation, play, and presence.

Hour at work. Picking an hour every morning at work to keep your phone off may be the way to start for you. Choose a time when you know it's okay to be unavailable or a time when you need to concentrate or get creative work done.

Hour for silence. You may choose the first or last hour of your day to turn your phone off. This can create meaningful space for solitude and silence. Consider using Do Not Disturb or a similar setting on your phones to set up regular phone-free hours. Better yet, in his excellent book on the topic, *The Tech-Wise Family*, Andy Crouch suggested putting your phone to bed before you go to bed, and waking up before your phone wakes up.

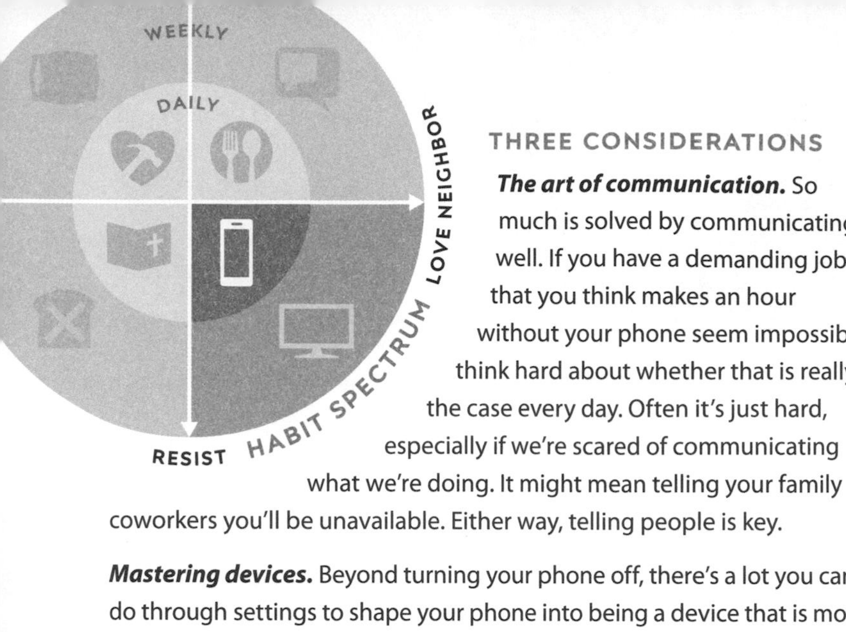

THREE CONSIDERATIONS

The art of communication. So much is solved by communicating well. If you have a demanding job that you think makes an hour without your phone seem impossible, think hard about whether that is really the case every day. Often it's just hard, especially if we're scared of communicating what we're doing. It might mean telling your family or coworkers you'll be unavailable. Either way, telling people is key.

Mastering devices. Beyond turning your phone off, there's a lot you can do through settings to shape your phone into being a device that is more likely to communicate your presence instead of divide your presence. I suggest turning off all notifications, and then over a week, turn the ones you truly miss or need back on. Also use voice controls whenever possible. This often will keep you from opening the phone and then getting distracted by ten things when you only needed to do one thing.

A place for phones. Whether at work or home, consider having a place for your phone. Set up a charger, put your phone there, and leave it there. At work I keep mine across the room, where I can't reach it or see it. At home I put it up on the mantle or in my dresser drawer. Consider having a charging station at the front door, so when your friends come over, you can offer a place where they can leave their phones. Having a place for phones goes a long way toward putting them in their place.

THE SMARTPHONE IS A TOOL THAT ENABLES MANY THINGS, BUT IT WILL NEVER MULTIPLY OUR PRESENCE.

READING AND RESOURCES

Alone Together, Sherry Turkle

Deep Work, Cal Newport

Irresistible, Adam Alter

The World Beyond Your Head, Matthew B. Crawford

DAILY HABIT 4

SCRIPTURE BEFORE PHONE

Now, in one hour's time, I'll be out there again. I'll raise my eyes and look down that corridor, four-feet wide, with ten lonely seconds to justify my whole existence. But will I? . . . I've known the fear of losing. But now I'm almost too frightened to win.
HAROLD ABRAHAMS IN *CHARIOTS OF FIRE*

I am not, in my natural state, nearly so much of a person as I like to believe: most of what I call "me" can be very easily explained. It is when I turn to Christ, when I give myself up to His Personality, that I first begin to have a real personality of my own.
C. S. LEWIS

LONDON CALLING

During my first year as a mergers and acquisitions attorney, I frequently worked with our London office, which was five hours ahead of the office in the United States. This meant that by the time I woke up every morning, they had already generated half a day's worth of emails for me to find waiting in my inbox.

Working with the London office was intense, but it was a good project for me as a new lawyer. We were trying to list an energy company on one of the junior markets on the London exchange, so I was learning about public markets, new energy technologies, and cross-border transactions all at the same time.

I wanted to do well, so I began trying to prioritize the London office's tasks each day. Thus began a new morning routine. I would wake up, roll over, and check my email to see what the London office needed. I was barely awake, but I was already evaluating whether I could get it all done that day and what needed to be pushed off to make it happen. In a way, it was almost like a first cup of coffee. It got my mind working and pulled me out of sleep.

Of course, after a couple of weeks of this, it became a habit. Without thinking, and even on days when there was a lull in the project, I would wake up to the blue glow of my inbox and the list of tasks to accomplish that day. I didn't notice what was happening on a deeper level until one morning, months later, I woke because one of my sons was crying. I grabbed my phone and opened it to flip through some work emails and try to get a grip on the day's tasks before I went to calm him.

Unfortunately there were more than I expected. A minute or two passed as I mentally cataloged them. I began typing a response on one. Then I felt an inexplicable stress rising. Was I worried about something?

That's when it hit me that my son was still crying while I was reading work emails that nobody—not even the people who sent them—expected me to respond to for hours. And yet something in me hungered to read and reply immediately.

How was it that I had come to a place where my heart was more responsive to the cries of my office than the cries of my son? It is because the moments of waking are powerful moments of formation.

Over the course of many months, my head was asking my phone a very practical question: *What do I need to do today?* But in the same moments, under the radar, my heart was asking my phone a much more profound question: *Who do I need to become today?*

WHO AM I? AND WHO AM I BECOMING?

Who am I? And who am I becoming? These are the questions our morning routines are inevitably asking and answering for us. But no words except the words of Scripture can bear the weight of a response to these questions.

The story of Scripture is clear. We do not know who we are apart from the God who made us, and we do not know who we are becoming apart from the God who is renewing us. We long to know who we are. We daydream about the versions of ourselves that we hope to become. But apart from Jesus we can do neither of these things.

Picture a broken mirror. Picture the shards littering the earth. That's what we humans are—broken reflectors. Alone we reflect a little piece of who God is. Together and redeemed by Christ, we will reflect him fully. But this means we can't know who we are by looking deep inside, discovering a true identity, and then becoming more like that person out of sheer willpower. That is not how it works *at all*.

When we look inside, we find a self in conflict, a mass of warring identities. This is because we are created in the image of God but fallen, so good and evil war within us. It's also because we are little mirrors made to reflect big things, but our internal sense of identity actually ends up reflecting the many different things we've spent our life looking at, whether good or not.

Consequently, we have a thousand different versions of ourselves all in disagreement with each other. There are versions of ourselves we think we should be, versions our spouse thinks we should be, versions our parents think we should be, versions our culture urges us to be—and the list goes on. What there is *not* is a version that is truly right because it is truly us. That is a myth—a popular one, but a myth all the same.

This means that the way we guide our formation is not by looking in and choosing our favorite identity; it is actually by looking out. We become what or who we reflect, which is to say we become what we pay attention to. We can't become ourselves by ourselves. The way we discover ourselves is by staring at someone else.

This can be dangerous. Staring at another broken shard of glass only makes us more broken. But when we turn our eyes toward Jesus, only there do we finally see the kind of person we were made to be like. We are children of the king, perfectly loved—not because that's "just who we are" but because that's who he is making us.

In his death and resurrection, Jesus gave us his place in the universe, heirs to the King of the new heavens and the new earth. Our most true sense of identity is found only in the story of who we are becoming, and that story is found in the words of the Bible.

We can become ourselves only by gazing on that story, but every morning there are other stories competing for our identity. The Common Rule habit of Scripture before phone is intended to cultivate the habit of resisting those stories and embracing the true story.

WORK EMAILS AND MORNING LITURGIES OF JUSTIFICATION

I often turn to work and career in the search for my identity. I hope that by achieving success I will finally become someone who is approved of. This means—as an identity reflector—I have to look at other people's faces, and my identity is tacked to whether they look happy with me and my work or not.

The habit of checking work emails first thing every morning encouraged this misguided search for identity because it started my day with the questions: *What do I need to do to make someone else happy with me? How can I justify my existence in the world today?*

After the morning when I read emails through my son's cries, I began to realize that something much more existential was happening in my morning routine, and I began to wonder if I should change it.

As a child, I grew up expecting to find my dad reading in his study every morning when I woke up. Usually it was his Bible. I'm sure he didn't do it *every* day, but I know that when I got up in the morning, I expected to see him at his desk with his Bible and notebook. I actually have his Bible on my desk right now as I'm writing.

On a random flip to open that Bible, I've landed in Colossians. There are a series of dates written across the chapters. On January 7, 2002, he read chapter one. On January 8, he read chapter two. And so on.

It amazes me that these were the months after his loss of the Virginia governor's race. He was a politician for some fifteen years of my upbringing, and that race, his biggest race, was the only one he ever lost. I remember crying myself to sleep that night (admittedly because I had really wanted to live in the Virginia governor's mansion and have my own bodyguard).

If ever there was a time for soul searching—*Who am I now?*—it seemed like losing a statewide election would be it. In politics, you literally stand up before the masses and say, "Approve of me! Choose me as your favorite!" If it were me, I don't think an election loss would have just destabilized my career; it would have destabilized who I was.

And yet first thing on the morning after that loss, he made us pancakes and told us about what he'd read in the Bible that morning. He was excited. He was wondering what was next for him. He was okay because he *knew who he was.*

I think my dad managed to keep a stable identity through a roller coaster of work success and failures because of the habit of looking each morning to God's love before he turned to look at the world. And once you know who you are in God, you can turn to the world in love. But if you don't, you'll turn to the world looking for love. So much of our identity hinges on this ordering.

THE NEWS AND MORNING LITURGIES OF ANGER AND FEAR

Shortly after that season at work with the London office, I began trying to resist checking work emails in bed. In my job—as is true of a multitude of jobs—you have to be responsive in order to hold up your end of the employment bargain. So it wasn't that I refused to look at my email in the morning; I just tried not to do it in bed. But I never decided what else I was going to do first.

This is where resistance needs to be paired with embrace. Resisting work emails in bed worked on its own, but the identity vacuum was quickly filled with something else—the news. Much of this was taking place in 2016, the most scandalous and tumultuous election season of my generation. On the national stage, the two presidential candidates were strangely similar. They both had remarkable wealth, a litany of personal scandals, and a disturbing ability to stir up pure hatred in their opponents. The media rejoiced. Nothing produced more dramatic headlines than these two people.

In the meantime, I was furious. There was so much coverage of them, both of whom were purporting to be good leaders. There was also an unceasing cry of doom and fear from all sides. Above all, I was most angry that for some it wouldn't matter all that much who got elected. We who are economically stable would figure out a way to preserve our interests. But for the most vulnerable among us—the unborn, the poor, single mothers, immigrants, refugees, minorities, prisoners—it would matter a lot. Unfortunately, it would happen in different ways depending on which candidate was elected. But the news wasn't about the people who would bear the brunt of our broken politics. It was about politicians as celebrities and the salacious details of their lives of sex, money, and power.

Here's the irony: I was so mad I could not wait to read the headlines when I woke up! I hungrily scrolled Twitter and news alerts. This became my new morning routine, and it solidified a new identity: I am the righteous judge who gets it—and nobody else gets it.

Anger and fear have something in common: we become the center of things. This is why so many of our conversations about headlines start with "Can you believe . . . ?" We're amazed and indignant that the world doesn't understand.

It's important to realize how natural and unnatural this is. It's natural because all people are generally scared and judgmental. That's what sin has made us to be. But it's unnatural because media companies prey on that. The news is tailored to incite anger and fear for a financial reason: Nothing brings us back for more headlines (and therefore ads) like anger and fear. They get rich, we get mad.

At one point that summer, I picked up a commentary on Isaiah that a former professor had given me long ago. I hadn't read Isaiah for years, so I decided to browse it. I had forgotten the historical context: two sides (countries in this case) battling to be Israel's savior while the Lord told them not to capitulate to either and that he would be their savior. Something resonated very deeply, even subconsciously.

I began to make Isaiah my morning readings, and I quickly fell headlong into a very, *very* different story of what was going on. It was a story of a God who loves and defends the poor and the vulnerable, a God who knows how to be righteously angry over injustice while remaining tender to both the victims *and* the perpetrators of that injustice.

Isaiah began to ease me away from being the one who is mad at those who don't get it. Isaiah made me doubt that I was the one who knew what to do, and it made me wonder if my justice compass needed to be re-tuned daily by the words of the prophets. Most of all, Isaiah rebuked my fear of doom while still giving voice to righteous anger. God will avenge all injustice. There's no question of that. That's why we can be at peace even when expressing righteous anger.

The questions of whether we let pundits or prophets calibrate our morning identity is an urgent matter of neighbor love. So long as we look to the news for our identity, we won't respond to the information with genuine care and concern for our neighbor. We'll respond by being indignant, a feeling alleviated by aligning ourselves with a tribe against a perceived wrong. All you have to do is pick a side; conveniently, that requires no repentance.

Aligning our identity to the king—over the country—is radically different. Only when we're secure in our identity as children of the coming King, who will right all wrongs, can we read the news for the sake of our neighbors' needs instead of for the sake of our own inadequacy. Only then are we able to repent, and not just blame another side.

When we're Christians first we can finally be good citizens second. That's the only way we can avoid being, in the words of Reverend William Sloane Coffin Jr., either uncritical lovers of country or loveless critics of country. But when we're citizens of heaven first, we finally become loving critics of country next—which is the truest kind of patriotism.

As a result of this new understanding, I decided it wasn't a good idea for me to read the news in bed. My options were beginning to narrow.

SOCIAL MEDIA AND MORNING LITURGIES OF ENVY AND VANITY

Social media has always been a great temptation for me because I'm such a vain person. I've never been able to sign on to social media without my mind and heart immediately beginning to spin: *How many followers or likes do I have? Why does so-and-so have more when she always posts stupid kid stuff? Also I know for a fact that her kids are nowhere near as calm and well-behaved as those pictures suggest. Wait, when did he get promoted? Is his new title better than mine?*

My identity spiral rages on. And for that reason, I've spent most of my life simply avoiding social media. Forgetting your password to Facebook and deleting Instagram are great ways to carry on life without the troubles of social media, which I have done for years. I was pleasantly aloof.

It was the height of irony, of course, that in initial calls with my publisher this conversation happened:

> Publisher: One of the important parts of marketing this book will be having an active social media presence.
>
> Me: Oh. [long pause] Like—um. How *active* do you mean when you say *active*?
>
> Publisher: Consistency is better than volume, but maybe a couple tweets a day and one or two Facebook posts a day.
>
> My silent thought bubble: *Have you even read what my book is about?*

I was easily convinced, of course, that some social media activity would be a great way to get the Common Rule habits into the lives of more people, but I was left with an uncomfortable reality: it is so much easier to judge the world when you remove yourself from the world. This was the way of holy safety I had been living for so long. Stay pure by staying out.

The monastic impulse in me said I should continue to stay away from social media, but the missional impulse in me said that's where the neighbors who need the love of God are. I want to acknowledge that both impulses have real elements of truth, many people at different times in life may appropriately choose one or the other, but my time had come. If I wanted to speak love to the world, social media would be one way to have the world's ears. I needed to learn to speak that language, just as I needed to learn Mandarin to be a missionary to the Chinese.

Many of my habits on social media were born out of this conflict, which I think is paradigmatic for the missional posture in any place, including us Americans who find ourselves as missionaries to America. I wonder, *How do we stay just far enough from the world to love it well? How do we be against the culture for the sake of culture? How do we be in these realms but not of these realms*? Like most questions in life, these require careful and communal thinking—and a lot of failure—before we find answers.

My friends helped me think through some simple practices that keep social media in its proper place. Here are some things I have learned: First, I try to open a media site only when I have need to post or respond. I don't open it because I'm bored or have a spare moment. Those spare moments are reserved for staring at walls, which is infinitely more useful. This is to say, I try to treat social media like work. I go to it once in the morning, once in the early afternoon, and once in the evening to put out content that

I think will help someone or to engage with someone who is responding in a healthy way.

Second, I avoid unplanned scrolling. Unplanned scrolling usually means I'm hungry for something to catch my eye—and plenty of strange, dark, and bizarre things are happy to catch the eye on social media. Planned scrolling can be very different. If you carefully curate *what* is in your feed and *when* you will scroll, the dynamic radically shifts. But in general, I believe we should be wary of the flicking thumb motion. The restless thumb often correlates to the restless heart.

Third, I turn off notifications. There is no good reason I (or any human being) needs to know in real time who is liking my posts when and how much. There are some useful purposes for these stats, but not as an every-moment affair.

Fourth, I don't use social media in bed. Beds are most useful for rest and sex (and sometimes reading a book). Social media is many things, but it is *not* a place of rest and *should not* be a place of sex—even though there are colossal temptations to use it for both. Mixing social media and bed tempts me to confuse these lines, and there is an easy way out of this unfair fight: *throw the phone out of the bed.*

Fifth, when I come across mean things said about me or someone I love, I employ the timeless strategy of any veteran parent: *ignore the temper tantrum.* Words are not nearly as useful as silence. Social media is a useful medium for some things but anger is not one of them.

I fail at all of these regularly, but they're still great rules of thumb. I still struggle with envy. I could list the people I know with more followers than I have. Sometimes I stand over the screen, hovering with a finger while I say to myself, *Do not open this just to check for likes.* Then I fail and do it.

But the more I use social media, the more I realize that the great danger is not in simply overusing social media, it is in *living through* social media. The problem is not so much the way it wastes time, it is the way it *frames* time. Without limits, we begin to see our whole life through it. We see our whole day through a possible post. We look around, wondering what in our field of view is worth taking a picture of. We listen to every conversation for a tweetable quote, instead of trying to understand the human being who is talking. We avoid disagreement in public, yet we express our most ardent emotions in carefully crafted Facebook replies or all-caps tweets.

This is no way to live. In fact, it's a miserable way to live. There is no love of neighbor in it, and there is no solution for it outside of becoming formed in the love story of Scripture.

If we wake every morning to social media, we *will* be formed in its lens on life and all the envy and self-righteousness that goes with that. But fortunately there's a different way. The Bible tells a story of us, not as people who were made to see and be seen or judge and be judged, but as children who were made to love and be loved. Only when we feel that in our bones can we use social media to love neighbors instead of trying to get their love.

BECOMING THE CHILD OF THE KING

There was something I didn't tell you about the page I randomly flipped my dad's Bible open to. Each of those dates had the annotation "w / Justin."

On January 7, 9, and 10—the very days I happened to open to—were mornings my dad and I spent together in the winter of 2002. I was a senior in high school, and every morning he would wake me and invite me to read the Bible with him before school.

I was exactly what you would expect in a highschooler then. I had a mass of warring identities. I was trying to figure out which friend, which party, which drinking game, which girl, which grade, and which band would finally give me the identity I needed in order to feel good about myself.

It was those days—precisely when I probably seemed the least receptive to parenting—that my dad was most vigilant about inviting me to his study in the morning to pray. I wanted to join because I knew he had the stable identity that I longed for, even if I didn't own that publicly. But so often I just couldn't get out of bed, and if I did, I nodded off during prayer. Fifteen years later, I can't recall a single passage of Scripture we read together.

But I do remember something. Something really important.

What I remember is having a dad who wanted to be with me. What I remember is reading together about a God who wants to be with us. Over the course of many mornings, by habit, my dad taught me that I was his, and that we were God's, and that no matter what went on in the world, we were both children, dearly loved.

Each morning presents us with these questions: Who am I? And who am I becoming? Each morning, the Scriptures answer the same, as God says, "You are my child, and you are becoming like me." That is something to stand the day on.

We can't become ourselves by looking inward, and we can't become ourselves by staring at our strange reflections in the screen. We have to look into the Word. Like the apostle Peter said, "To whom shall we go? You have the words of life" (John 6:68).

Cultivating the habit of Scripture before phone means looking in the right place to ask who you are. You open the Bible, and you find you are with your dad. You find your name written on its pages. You find you are loved. Then you begin to reflect that love, just as you were made to.

DAILY HABIT 4
SCRIPTURE BEFORE PHONE

THE HABIT AT A GLANCE

Refusing to check the phone until after reading a passage of Scripture is a way of replacing the question "What do I need to do today?" with a better one, "Who am I and who am I becoming?" We have no stable identity outside of Jesus. Daily immersion in the Scriptures resists the anxiety of emails, the anger of news, and the envy of social media. Instead it forms us daily in our true identity as children of the King, dearly loved.

THREE WAYS TO START

Reading plans. Getting a daily devotional or a book of liturgical readings such as *The Book of Common Prayer* is a great way to start these daily readings. If you don't have one of these, consider trying a month with the following:

- *Psalms.* Whether you go through them in order or otherwise, reading a morning psalm is always a great place to start.
- *Matthew.* The Gospel of Matthew has twenty-eight chapters; try reading one each morning for a month.
- *Romans.* The book of Romans has sixteen chapters; try reading half a chapter each morning for a month.

Daily prayer apps. I prefer reading out of a paper Bible because of the way tactile engagement brings my mind to focus. However, I often find a Bible or prayer app on my phone very useful for morning readings, especially if I'm traveling or if I have to get out of the house early. The Daily Prayer app is a great place to start. The ESV Bible app also has great reading plan functions.

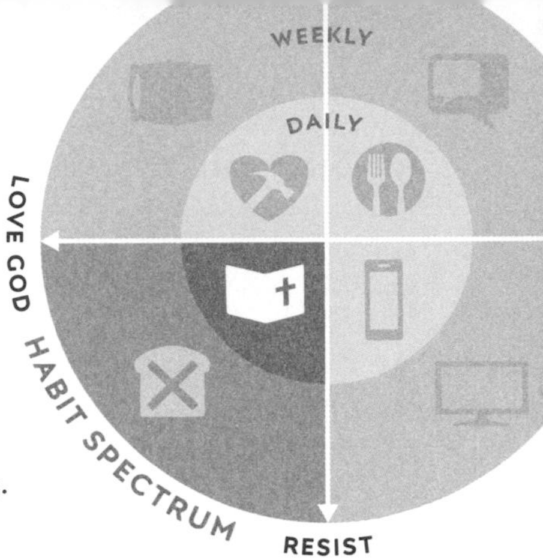

Creating a new routine. The best use of this habit is simply to get your phone out of your morning routine entirely. Try starting with a week and leaving your phone alone for the first hour of each morning. Do coffee and Scripture and then add journaling, meditating, other readings, or exercising.

THREE CONSIDERATIONS

Seasons. Many of us have jobs that require attention first thing in the morning. I have seasons like that, when a project is important enough or around-the-clock enough that I need to check on things early each morning. (If you think that is every day for your life, gently reevaluate your understanding of your importance.) If you're in a season when you must, I've been there. I usually try to read a psalm then check email. If there is something, that becomes my morning. If not, I put my phone back upstairs and continue with my desired morning routine.

Long study. Every follower of Jesus should be studying the Bible in depth. But if you're not a vocational minister or on a weekend retreat, you're likely not going to do it every single day, and that's okay. Let longer times of reading a book or commentary, or making an in-depth annotation of a passage, grow out of reading your Bible regularly (not get in the way of it). Remember, these habits build a trellis that amazing new things grow on. Short daily readings don't undercut longer study, they build the foundation for it.

Journaling. Journaling is a keystone habit—that is, it changes everything else in your life. If you make a habit of filling up one blank page while you read or pray or are silent before picking up your phone, your life will change.

> THE STORY OF SCRIPTURE IS CLEAR THAT WE DO NOT KNOW WHO WE ARE APART FROM THE GOD WHO MADE US, AND WE DO NOT KNOW WHO WE ARE BECOMING APART FROM THE GOD WHO IS RENEWING US.

READING AND RESOURCES

The Book of Common Prayer

God's Wisdom for Navigating Life: A Year of Daily Devotions in the Book of Proverbs, Kathy Keller and Tim Keller

ONE HOUR OF CONVERSATION WITH A FRIEND

Without communication, there can be no community. . . . That is why conversation, discussion, or talk is the most important form of speaking or listening.
MORTIMER J. ADLER

"No longer do I call you servants, for the servant does not know what his master is doing; but I have called you friends, for all that I have heard from my Father I have made known to you."
JESUS (JOHN 15:15)

Whenever you go out, walk together, and when you reach your destination, stay together.
THE RULE OF ST. AUGUSTINE

BEING KNOWN

It was fall, and I was sitting in my living room with a close friend when I got a call with some bad news. Someone we both knew had become badly addicted to prescription drugs. This was scary, not just because we hear about these kinds of problems often, but because people seem so "normal" before you learn their terrible secrets. We all seem to have a capacity to make ourselves appear okay while hiding something that slowly kills us from the inside out.

As we sat there in our shock and sadness, it's worth noting what we *did not* say. We did not say, "How could this happen?" We did not say, "What are the gory details?" We did not say, "How lucky we are, not to struggle with something like this."

Something much more personal floated about the room. It was an unspoken question: *Is there anything you aren't telling me?*

That is the question that goes either unasked or unanswered in so many lives that collapse. Often, if honestly asked and honestly answered, this question can turn whole lives around. Secrets feed off going untold, and darkness exists only where there is no light.

So that's what we asked each other. As if the world were on pause, in a quiet living room on a cool fall evening with the kids sleeping upstairs, the question was asked and answered. And the answer we both gave was "No. You know everything."

The weekly habit of an hour of conversation is meant to cultivate this kind of life, where you know and are known by those closest to you.

MADE FOR FRIENDSHIP

The question "Is there anything you aren't telling me?" gets at the heart of friendship, because friendship is being known by someone else and loved anyway. Friendships in which we're vulnerable make or break our lives. With them we thrive, and without them an essential part of us—if not all of us—dies.

One of the defining marks of the Christian faith is that God is three persons in one triune God. Among the thousands of radical implications of the Trinity, my favorite is that God is a fellowship. That means we are made in the image of fellowship.

This tells a very different story of who we are than what we hear and see. This means we didn't come from chaos, as if we are simply accidental life going from random darkness to darkness. We did not come forth from loneliness, as if some needy god needed lesser beings to sing its praises. We came from *friendship*.

Everything in the universe has its roots in friendship. That means the longing to be in right relationship with other people and things is at the heart of every molecule in existence—and most powerfully in our own hearts.

We can't be happy without knowing and being known, because that's the image of trinitarian friendship we were made in. This explains why there is only one time in the creation story when God says the words "not good," and it is when a man is alone (Genesis 2:18). Everything else that comes before is pronounced "good." God looked at the molecule of twin hydrogen plus oxygen—*Good!* The arctic coastline—*Good!* The acai palm tree—*Good!* The antelope—*Good!*

But something was *not good* about the pinnacle of all creation—man. The Lord saw that it was not good for man to be alone. It would seem that God is all we need, but Genesis tells us that because God made us in the image of himself, we are created for relationship. Thus, even in the Garden of Eden, Adam was lonely until God gave him Eve. So when Adam saw Eve, he sprang into song, "bone of my bone and flesh of my flesh" (Genesis 2:23). This was not just an ode to man and woman. It is also an ode to friendship.

Adam saw that he was made for this, which means that simply to be happy in the presence of a friend is like a song or a prayer to God's glory and generous design. God loves it when we embody friendship, because when we do that, we are embodying the triune God himself, and we feel his pleasure.

CONVERSATION AS FRIENDSHIP

My family moved from Chesapeake, Virginia, to Richmond, Virginia, the summer before I started high school, and I remember the fear of watching my old driveway recede in the distance. I lived next door to my best friend, and now all that was over. High school in a new city began with the terrible anxieties of loneliness and fear that plague all of us when we don't have a friend who walks out into the world beside us.

For that reason, I can still remember the moment in tenth grade when I met my then (and still now) best friend, Steve. It was exactly what C. S. Lewis described as the moment that friendships are born "when one person says to another: 'What! You too? I thought I was the only one.'"[1] We both skateboarded. We both played the drums. We both projected a false confidence to hide our deep insecurities. We both desperately wanted to be popular and went to the same youth group. In the best and worst of things, we found a seed of friendship in the mutual exclamation, "You too?"

But these things were only the seeds of friendship. I remember sitting by campfires, talking late into the night. I remember long road trips to see concerts, where we talked a lot. I remember lying on floors in sleeping bags, talking after everyone else had fallen asleep. I have tremendous memories of many adventures and misadventures with many people in high school, but the reason Steve and I became inseparably close was because, beyond the shared activities, we shared words. Without the work of real conversation, where your deepest hopes are admitted and your greatest secrets are discovered, relationships remain the mere common interests of acquaintances.

Vulnerability and time turn people who have a relationship into people who have a friendship. That's what friendship is: vulnerability across time. The practice of conversation is the basis of friendship because it's in conversation that we become exposed to each other.

This is not to set aside the importance of shared identities, shared activities, shared concerns, and all the other things that spark friendship. This is only to say that all of the above find their consummation in being savored through talk. In conversation we disclose our loves for such things and, by doing so, disclose our very selves.

> VULNERABILITY AND TIME TURN PEOPLE WHO HAVE A RELATIONSHIP INTO PEOPLE WHO HAVE A FRIENDSHIP. THAT'S WHAT FRIENDSHIP IS: VULNERABILITY ACROSS TIME.

Conversation exposes us in two ways. First, face-to-face conversation brings risk. Face to face in real time, we're less predictable and less guarded. In her book *Reclaiming Conversation*, Sherry Turkle describes the way texting and online chatting have threatened true friendship because they allow us to plan and curate the versions of ourselves that we bring to our discussions. When we're removed from facial expressions, body language, and tone of voice, and when we have time to consider and edit our replies, we don't face the risk that face-to-face conversation naturally brings. So we don't risk being known as someone less than perfect.

Turkle writes of interviews with college students of the digital age who find "unrehearsed real-time conversation is something that makes you 'unnecessarily' vulnerable."[2] But to be vulnerable is precisely the point of conversation, because in vulnerability we are finally truly known.

The second way conversation exposes us is by the truth-telling that happens in it. This goes beyond the risk of being face to face and further into the risk of the question my friend and I asked in the living room that night: Is there anything you aren't telling me?

There's nothing more terrifying and redemptive than removing the fig leaf and telling who you are to a friend. It's terrifying because we are never who we wish we were. It's redemptive because that's at the core of enacting the gospel in communal life.

Friendships embody the power of the gospel in a unique way, because in friendship we live out the truth of the gospel to each other. What is the gospel besides that Jesus knows how broken we are and sticks around to love us anyway? What is a friend except someone who knows how broken we are yet sticks around to love us anyway?

Conversation is the beginning of all of this, because it is the place we become truly known. Looking back to the year when Steve and I became friends, I see the real blossom of friendship between him and me. It wasn't in common interests but in common conversation. It was when we began to tell each other the things we wouldn't tell anyone else: what we wanted to do with our lives, who we wanted to become, who we liked, and what we were scared of.

Much of friendship comes from admitting the things that make you seem fragile when spoken out loud. And this is why friendship is so hard. Vulnerability is risky, and time is limited.

How do we create a life of friendship when we have neither the courage nor the time to talk? The answer is to practice courage and prioritize time.

We need the courage to be vulnerable in a world that is scared, and we need to make it a priority to take time in a world that is distracted. The habit of a weekly hour in conversation can cultivate both.

THE POWER OF VULNERABILITY

Go back to the moment I mentioned at the outset, about my friend and me in the living room, asking each other, "Is there anything you aren't telling me?" Well, the next night, there was a knock on the door. I opened the door, and that same friend was standing on the porch. Almost no one just knocks anymore, so when I opened the door and saw him standing silently on the front stoop, I knew something was wrong.

"We need to talk," he said. Lauren was still awake, so I poured us two glasses of something to sip and suggested that we sit on the porch.

I'll never forget the combination of fear and courage in his eyes when he said, "You remember the question we asked last night, about anything we haven't told each other?"

I knew where this was going. I knew it because he had a look in his eyes that caused me to remember a feeling ten years before, sitting with my dad in a restaurant when I was in college, waiting for the moment when I could work up the courage to tell him I had been lying to him—and lying for a long time. It's worth noting that as hard as telling the truth is, telling someone that you *haven't* been telling the truth is even harder.

That's what my friend was doing. There *was* something, after all—something he was holding back. He had lied, just the night before. Boldly and to my face. But now, with *even more boldness*, he had come back to tell the truth.

What he said, of course, is just between me and him. But it was plenty serious, a struggle deep and dark enough that if not exposed to the light of conversation it might have overtaken him.

In retrospect, it was hard to tell which one of us was more broken up about it—him for the guilt or me for the feeling of being shut out. That night I leaned on my dresser and wept, openly and uncontrollably. I wept and I wept and I wept. In part it was because of the lie; lies always threaten the lifeblood of relationship. But even more so it was because of the threat of the evil. So often I'm tempted to think that life is not, after all, very high stakes. I'm tempted to believe the fiction that evil is more like a cat that I can bat away than a lion that is "crouching at the door," waiting to pounce on us (see Genesis 4:7 and 1 Peter 5:8).

There is a pain to the realization that no matter how hard we try, my friends and I are not immune to the dangerous addictions that come from the temptations and exhaustions of life. We aren't immune to the uncontrollable anger that comes from pain. We aren't immune to the dangerous lies that come from the fear of life itself.

When I was barely twenty and envisioning the future for me and my friends, I had no idea how difficult a life lived alongside each other would be. I had no idea that before I was thirty, I would have friends look me in the eye and say, "I don't believe in this Jesus anymore." I had no idea I would have friends that would call me and say what happened at a nightclub the night before. I had no idea I would have friends sit me down and confess addictions to alcohol, prescription drugs, pornography, or worse. Perhaps more importantly, I had no idea that I would be calling them to confess that at the age of twenty-nine I needed pills or alcohol just to fall asleep.

My coming of age has meant coming to grips with the startling evil in the world. In Genesis, God told Cain that sin was crouching at his door, wanting to rule over him (Genesis 4:7). I used to think that story was about someone else. Now I see that it's about all of us. I read it and I see myself, and I see the people I love most. I see the world as a place where lions prowl, desperate to tear you apart. But inevitably the most vicious lion is within. He is born as a little secret that no one needs to know, and he grows into a monster that rips you apart from the inside out.

This is the truth of the darkness within. But the darkness is never as strong as the light.

Here is the power that lies in vulnerable friendships: *together we beat back the darkness by exposing it to light*. When the darkness of our lives is exposed by the light of vulnerable conversation, we participate in opening our darkness to the power of the gospel. That night in the restaurant when I confessed to my dad and he forgave me, my whole life turned around. That night out on the porch after that friend came back to tell the truth, his life completely pivoted.

Did you notice the common thread in all the things I just mentioned that I never expected I would hear from my friends? *It is that they have been spoken*. They have been brought to light. That one friend believes in Jesus again. The other quit the nightclubs. The addicts are clean. I've learned to drink responsibly, even on the evenings I can't fall asleep, which are now—praise God—almost never.

The darkness rages in us, but honest conversation is a practice of light. And the incredible thing about light in the dark is that the light always win.

THE POWER OF TIME

The vulnerable friendships that embody the gospel don't happen because we wish we had them; they happen because they're cultivated over time. They grow because we arrange the trellis of habit that allows them to flourish.

Friendships are hard when you don't actually have time together, which is why friendships are not just about vulnerability but also about time. While Lauren and I lived in China as missionaries and then in Washington, DC, during my law school years, most of my friends were migrating back to Richmond. It's an amazing city, but most of them were moving there not for the city but for each other.

When I started law school, Lauren and I imagined we would move to wherever the most prestigious job was. I figured it would be a move to New York or maybe back to Shanghai. But a nagging question kept coming back, especially as having our first two kids narrowed our free time immensely: Why live far away from your friends, unless there's a really good reason to?

Put more generally, why do we arrange our geography and our schedules in a way that makes putting consistent time into friendships so hard? For me, the reason was career. I wanted the best job, but I began to wonder whether a life with the best job was worth a life without a best friend.

The Lord had called me to China and called me to law school; there was no doubt about either of those in my mind. They were good reasons to be away from my friends. But toward the end of law school, I didn't hear God calling me anyplace instead. Absent that calling, Lauren and I began to consider moving for friends instead of for jobs. I felt a clear call to arrange my life for the sake of friendships, so we decided to commit to Richmond, where our friends were, and to try to find a job from there.

My second son was born the month we moved. Because I was experiencing a deep sense of the importance of friendship, we named him Asher Stephen Matthew Earley. *Asher* means "the happiness of God" or "the blessing of God," and Steve and Matt are my two best friends. They are also the names of Lauren's brother and a close family friend. We liked the way it blurs the lines of friends who feel like family and family who feel like friends. Asher was thus named as a praise and a prayer: a praise that we had found the happiness of God in friendship and a prayer that the same would be true for him.

I am deeply grateful that I live in a city where my friends live, but it's far from perfect. Just ask them. We all have kids now, we are deep into careers, and we meet new friends and have new obligations, so the time that friendships require becomes exactly what is so hard to give. We often feel spread impossibly thin.

That's why cultivating habits of devoting time to friendships is so important. The world would have us cultivate something else. The usual life in America leans toward busying yourself with things that seem urgent, but friendships will never seem urgent. The most important things never are until it's too late.

The Common Rule habit of a weekly hour of conversation is aimed directly at that struggle. The idea is to cultivate a keystone habit of being the vulnerable, relational person you were created to be, even when life becomes complicated. In the most demanding of times—when you have a newborn, a death in the family, a crushing work project, an aging parent—we tend to sacrifice times with friends, and yet friends are precisely what we need the most.

My friends cultivate weekly touch points for conversation in all kinds of ways. For me and Steve, it's a standing coffee every week. Some of my friends meet in groups of threes for dinner or breakfast weekly. Some have a community group where everyone meets every week. Some couples alternate weeks when the guys meet and talk and then the women meet and talk.

There are as many ways to do this as there are conversations to have. The point is that as our lives have grown more complicated with work and kids, these weekly touch points of open conversation sustain the lifeblood of our friendships across time.

Without them, we would be people who used to have friends. With them, we *are* friends.

THE OPEN CIRCLE OF TRINITARIAN FRIENDSHIP

It was my junior year of high school, and Steve and I were just about to jump into a friend's convertible to drive downtown when someone else showed up and asked to join. He was a freshman, and we didn't know him that well, but we couldn't think of a good reason to say no.

It became clear along the way that he had good taste in music and an even better sense of humor, but for most of the ride, we weren't quite sure how he'd made his way into our car. We didn't even know his name.

Regardless, it was clear he just wanted a friend. But Steve and I weren't willing to open up.

Like sex, food, work, technology, and everything else that God made good, the better something is, the more it can be twisted. One of the darkest twists on friendship is our tendency to ruin it by making it exclusive. As most of us know, there are few deeper pains than the feeling of being shut out of friendship. Unfortunately, that's what Steve and I did. We saw our friendship as something to protect, not something to risk for the sake of an outsider.

I remember that season being painful, for everyone. For this new guy, because he was just like us, someone looking for a friend and we initially denied him that. And it was painful for Steve and me because any blessing that you try to hoard for yourself begins to sour. We ruin the goodness of blessings when we refuse to use them to bless others.

I think it's especially important to see that keeping friendships closed is a broken twist on what true friendship is. The fundamental truth of friendships is not that love is limited but that love is infinite. We know this because the friendship of the Trinity did not generate less love but *more* love. By virtue of making us like him, God in creation *expanded* the circle of friends. Jesus now calls us his friends, and by saving us, he invites us *into* the dance of the Trinity. The circle of love is open and expanding.

The nature of true friendships is not to shut the outsider out, it is to draw them in.

As Steve and I continued to spend time and share conversation with this new friend, despite our selfish intentions to keep him at arm's length, we began to see that the fire burned bigger and brighter for all of us with him in it. We had new moments of saying "You too?" As our life of common interests turned again into a life of common conversations, this new friend got to see all the good and the bad in us. Most particularly, he got to see the ways we did and did not live up to the faith we claimed to have. This is not a shiny story: on the way to all becoming the best of friends, we hurt each other a lot. We made big mistakes.

This is, of course, the truth of looking closely into anyone's life—followers of Jesus included. What amazes you is how there is simultaneously so much beauty and darkness right alongside each other. The only explanation is our fallenness in light of Jesus' redemption. Praise God, grace prevailed, and these friendships grew.

This new friend got to see all of this, and after seeing the best and the worst of us, he was attracted enough to Jesus and disappointed enough in us to look beyond us to the Jesus that we worshiped. After much fun and sadness, this friend was baptized. His name is Matt, the Matthew of my son Asher's name.

BEFRIENDING THE WORLD

As I write this page in a small writing cabin in Nelson County, Virginia, Matt has packed up his family and all their belongings. They are driving up to Virginia from Florida. After a decade in the United States Navy, he is coming home to live with us in Richmond.

Tomorrow night I will go see him. I have bought some good wine. I will meet his three-month-old daughter. We will be glad, and we will have conversation. It has been fifteen years since we all lived in the same city, but here we are again. Soon he will be joining the weekly morning coffees with me and Steve.

Matt's friendship is a continual reminder that the fire of friendship is contagious. Opening outward is the truest direction of friendship. The circle grows. Here one plus one equals three or even four. The circle is complete, but it is somehow still open. Love defies mathematics and geometry.

If friendship is a practice that reminds us of what the gospel is, it is also a practice that puts the gospel on display to the world. In a culture of loneliness and individualism, there is no better witness to the Trinity than embodying a counterculture of real friendship.

Here then is another habit of Madeleine L'Engle's "light so lovely." Friendships light up the darkness. For that reason, I think of friendships as little fires we tend. They light up the truth of the gospel, they invite people into the warmth, and they become the fires around which many can gather.

To cultivate the practice of a weekly hour of conversation is to keep the fire burning. It is to look out into a cold and dark world and to offer some light, some warmth, and a place to sit and talk.

WEEKLY HABIT 1

ONE HOUR OF CONVERSATION WITH A FRIEND

THE HABIT AT A GLANCE

We were made for each other, and we can't become lovers of God and neighbor without intimate relationships where vulnerability is sustained across time. In habitual, face-to-face conversation with each other, we find a gospel practice; we are laid bare to each other and loved anyway.

THREE WAYS TO START

Standing meeting. Try setting up a standing time with a friend—such as every Thursday evening or every Friday morning—when you always get together. Don't be discouraged by the fact that sometimes you have to miss; be encouraged by the fact that the rule is getting together, and the exception is missing it sometimes.

For couples. Lauren and I have gone in and out of seasons with our friends Steve and Lindsay when we either get together as couples each week to talk or send the guys to one house and the girls to another to avoid the need for babysitting.

From roommates to friendship. Setting up a weekly time to eat and talk or share a drink can be the keystone habit that moves people from just roommates to real friends. Consider setting up a time when the people you live with get together not to talk about housekeeping but just to talk about life with no other distractions.

THREE CONSIDERATIONS

On telling secrets. The question is not how to tell secrets; everyone knows how. The question is whether it's worth it. With someone you trust and who loves you, it always is. Tell your secrets. Do it tonight. It

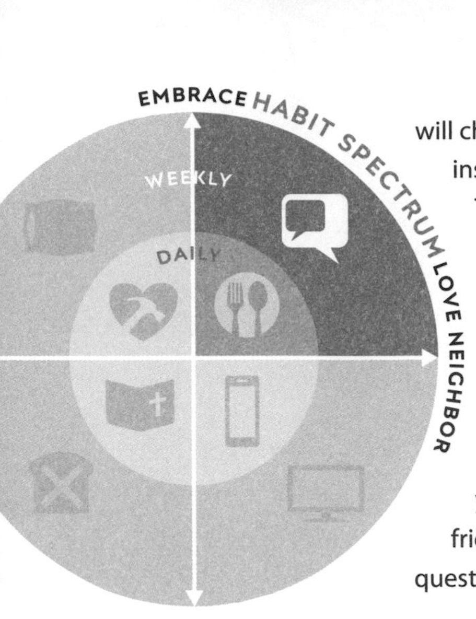

will change your life and will probably inspire your friends to tell theirs too. There's no bigger catalyst for deep relationships than telling your secrets.

The power of good questions. Often great conversations come from someone who has mastered the art of good questions. If that isn't you, consider brainstorming with your friends to come up with some good questions you can regularly ask each other.

Open friendships. While the habit of weekly conversation begins in intimacy, it ends in openness. The goal of gospel friendships is not just to enjoy the fruits of friendship together, but also to offer them as nourishment to the world. Think of your spaces of friendship as one of the first places to invite a new person you meet to.

THE DARKNESS RAGES IN US, BUT HONEST CONVERSATION IS A PRACTICE OF LIGHT.

READING AND RESOURCES

Spiritual Friendship, Wesley Hill

Made for Friendship, Drew Hunter and Ray Ortlund

WEEKLY HABIT 2

CURATE MEDIA TO FOUR HOURS

We write to find the fire in the darkness.
JOHN GREEN

Art is . . . a game played against chaos and death, against entropy. . . . [It] asserts and reasserts those values which hold off dissolution . . . rediscovers, generation by generation, what is necessary to humanness.
JOHN GARDNER, *ON MORAL FICTION*

The medium is the message.
MARSHALL McLUHAN

I was standing amid floor-to-ceiling shelves of books in wonder and awe when my view of stories suddenly and forever changed. There were enormous piles of books lying in corners. Books covered the walls. Books even lined the staircases as you went up from one floor to the next. It was as if this used bookstore was not just a place for selling used books; it was like the infrastructure itself was made up of books. There were books to hold more books, stories built out of stories.

I was standing in Daedalus Books in Charlottesville, Virginia, and I had recently read Mortimer J. Adler's *How to Read a Book*. I was alive with the desire to read. But at that particular moment, my glee turned to horror. For whatever reason, the truth of the numbers suddenly hit me. The year before, I had read about thirty books. For me, that was a new record. But then I started counting. I was in my early twenties, and with any luck I'd live at least fifty more years. At that rate, I'd have about 1,500 books in me, give or take.

There were more books than that on the single wall I was staring at.

That's when I had a realization of my mortality. My desire outpaced reality. I simply didn't have the life to read what I wanted to read.

Suddenly my choices in that bookstore became a profound act of deciding. The Latin root of the word *decide*—*cise* or *cide*—is to "cut off" or "kill." The idea is that to choose anything means to kill off other options you might have otherwise chosen. That day I realized that by choosing one story, I would have to cut off other stories. I had to choose one thing at the expense of many, many other things. I would have to choose carefully. I would have to curate my stories.

In the years since my revelation at Daedalus Books, which was barely ten years ago, the way that we consume stories has radically changed. The smartphone had not been released to the public. Companies such as Amazon and Netflix were still mailing books and DVDs. The "streaming story," as I like to call it, was not yet a reality.

Now, in less than a decade since that experience, technological and cultural shifts have collided in such a way that stories no longer sit on shelves and wait for us to pick them. They come at us. They pick us—sometimes in rapid and aggressive streams of media.

Curating stories used to be a matter of luxury. Now it's a matter of necessity—and perhaps even urgency.

STORIES AS FORMATION

We become the stories we consume. Stories make up our lives the way they made up the walls of that bookstore. This is not true by happenstance; it's true because we were made to live in a story.

Our story begins with the creation of the world. Eden is the opening scene of a movie in which everything is the way it's supposed to be. And then we're hooked by tragedy: the fall of man and the separation of humans from God. Then there's the chase, the conflict. How will God rescue his people? He shows his love for them in the Old Testament's dramatic scenes of romance, in tracking them down through deserts, and in saving them from war.

And then the plot thickens. Just when things couldn't get worse, we find out that God will not just come *for* his people; he will *become* his people. Jesus is the archetypal hero who comes in to save the world. But he won't save the world by fighting; he will save it by dying. The gospel is tragedy: Humanity will be saved, but God has to die.

But there is a twist. The resurrection defies all expectations. Evil is defeated, and good will reign. God will win after all, and love will prevail. There is a new time coming, and the kingdom will have no end. The gospel moves—in the words of Frederick Buechner—from tragedy, to comedy, to fairy tale.

The point is this: We don't just watch stories, we live in one.

We are characters in the most epic narrative of all time, and it is real. It is a story unfolding in actual time, and the stories we watch are all trying to explain to us what this real story is about. They help us figure out how to live in our own story.

When the theologian Stanley Hauerwas argued that story is "the necessary grammar of Christian convictions," he was saying that the Bible isn't primarily a moral message on what is right and wrong; it's a story of how God is saving us.[1] We can't understand morality from situation to situation unless we first understand the story we're in, who the hero is, how we are being rescued, and from what.

In *On Moral Fiction,* John Gardner wrote that a good story beats back the chaos of the world. A good story shows us that there is some kind of world with beginnings and endings, where things happen—and happen for a reason. Stories convince us that things are, after all, going somewhere.

That's why stories—more than any ethics lecture or Sunday school lesson—shape our idea of what the good life is, where the world is going, and what it means to be human. They tell us what is beautiful, what is just, and how we should live with each other.

This is why we don't just watch stories. We become them.

TECHNOLOGY AND THE ADVENT OF THE STREAMING STORY

This is nothing new. The formational power of stories is an ancient truth.

What is completely new, however, is the medium stories are now packaged in. In the span of less than a decade, the streaming story has completely reshaped the landscape of how stories come at us, each with the purpose of forming us through its plotline.

Technology has always been intertwined with story. In ancient times, to hear a story, there had to be a teller. You had to have a human—a bard—who memorized stories to tell or sing to you. The amphitheater was an early example of using the technology of architecture and space to amplify sound so more people could hear stories. Technologies such as paper and ink began to make stories less tied to the teller and accessible to those who could read. Then the printing press radically altered the number of stories that one could come across in one's lifetime.

Centuries later, as technology improved and literacy rates rose, new ways of storytelling began. Writers like Charles Dickens were among the first to produce mass-disseminated stories. For the first time in history, a large community of people collectively began to hang on a plotline.

Then technologies shifted again. Thomas Edison began to put pictures beside each other, and the Warner brothers bet a business on the idea of telling a story through moving pictures. The world was forever changed, of course.

The movies became a place to go, and then VCRs and DVDs brought them into homes. Suddenly, as if the past century hadn't seen enough change in the art of the story, the internet and the smartphone exponentially increased the amounts and kinds of stories we can consume in any given moment.

The fundamental power of the story and the desire to hear one is ancient. But the medium through which a story gets to us has always been intertwined with technology, and the current generation has experienced an unprecedented shift in the rise of the streaming story.

The "streaming story" is not a technical term. I made it up because I'm trying to talk about what it means that audio and video content are not only continually accessible but also targeted to find us even when we're trying to give our attention to something else. Stories are not only far more accessible than ever before, they are also far more invasive, as the writers often have all kinds of ulterior motives.

So when I use the term streaming story, I mean audio and video, short or long. I mean stories that come into our field of attention whether or not we want them. This is not only Netflix and YouTube. It is also NPR, Fox News, Amazon Prime, and the *New York Times*. This is the walls of sports bars and the counters of coffee shops. Increasingly, it is every other app we download. Instagram calls one of their feeds "stories" because it shares short, disappearing vignettes of our lives. Pornography sells its depravity not just in taboo images, but with videos that connote narratives, identities, and possibilities that capture more than just our physical lust—they capture our deeper desires to lord over or be lorded over. The streaming story is everywhere.

We're seeing a renaissance in both ultra-epic narratives that span hours of television, such as *Game of Thrones* and *Lost*, and also in micro stories that are seconds long but come into our field of view—usually as advertisements. A marketing firm here in Richmond has won awards for its four-second ads for GEICO, which pop up before you start a YouTube video, draw you into a seconds-long story, and then—*pow,* the big block GEICO letters come up.

Stories now press in on us much like the books at Daedalus Books. The walls of the world seem to be made of stories. And they come for us whether we want them or not.

When the media critic Marshall McLuhan wrote that "the medium is the message" in 1964, he was talking about the advent of TV and how rapidly it was changing communication forms. This has only become more true. Now the content of a story is not our only concern; the medium is equally *if not more* formative. The *way* we watch is as important as *what* we watch.

In the bookstore that day, I realized that to live well I would need to curate my stories as a consequence of my own limitations. But no books have ever come charging at me. It was hard enough to curate when I had to go to them.

Now, however, we don't choose our stories nearly as much as they choose us. Should we do nothing, someone else's stories will curate our lives for us. If we don't cut off their options; they will cut off our options.

And if stories are as formational as the Bible, and John Gardner and common experience would tell us they are, this now means that we live in a world of competing types of formation, streaming like busted faucets everywhere we look. We are guaranteed to be formed in consumption unless we ruthlessly pursue curation.

THE NEW VIRTUE OF CURATING STORIES

The new world of the streaming story means that we must urgently cultivate the virtue of curation. The four-hour part of this habit is admittedly arbitrary. You might pick two or you might pick twenty. The point is picking some limit that forces curation. You can't watch or listen to everything that everyone else is; you can't even watch or listen to everything you want to. But you can and should watch some things, even many things. But you also must curate them, lest someone else who does not love you curates them for you.

This habit of curating media intake strikes at the heart impulse of the Common Rule. The good life doesn't come from the ability to choose anything and everything; the good life comes from the ability to choose good things by setting limits.

Limits are where freedom is found. We don't need unlimited choices; that actually limits our ability to choose well. We need a limit on our choices, which actually empowers us to choose well. By limiting stories to a certain number of hours in a week, you introduce the ability to choose them well.

Curation implies a sense of the good. An art gallery has limited space on the wall, so its curator creates shows to make the best use of that space according to a vision for good art. I suggest we have a vision for good stories, and we curate accordingly. We could start with curating for beauty, justice, and community.

Curating for beauty. When I lived in China, one of my favorite places to go was a modern Chinese pottery shop. It had bare concrete walls that held heavy wooden shelves. The minimalist aesthetics made the pieces of fine white China pottery on those shelves stand out from the sparse wood and concrete like sudden sounds.

The first time I walked into that shop, I tiptoed around for an hour just looking at the pieces—quietly, as if not to disturb them. Many of the designs were inspired by nature. They had perfect functionality, as a bowl or mug, for example, while retaining a "this is something else too" feeling: a dessert plate as textured as a maple leaf or a ceramic teapot that somehow looked as soft as a peach.

I always had a strange feeling as I walked out, like some unknown hunger in me was satisfied, like my soul was full. The English language affords words for this experience. We say things like "feast your eyes on this," because we intuitively acknowledge we are hungry for beauty.

It wasn't until reading Genesis one day that I finally came to a theological understanding of what had been happening in the pottery shop all along. "And out of the ground the LORD God made to spring up every tree that is pleasant to the sight and good for food" (Genesis 2:9). This verse caught my eye because it explains the very metaphor that I had adopted to describe my pottery trips: sight and food. The stomach was made to hunger for food; the eye was made to hunger for beauty. We were made to consume beautiful things. Excellent music, great films, stunning performances—these are all food for the hungry soul.

Curating for beauty means moving past the idea that we must choose stories primarily based on their message and embrace the idea that we need beauty, full stop. And that beauty may be found in unlikely places.

Lauren and I love watching movies, but with my demanding job and four little boys, it's both unlikely and unadvisable that we try to watch every worthwhile movie recommended to us. Consequently Lauren listens to a film podcast called "Filmspotting" and reads numerous critics she trusts. She then tells me which movies are either so excellently made (signpost of beauty) they can't be ignored or such crowd-pleasing tales (signpost of universal truth) that everyone should watch them.

Sometimes these are best-picture nominees, and sometimes they're movies only the most obscure festivals have heard of. But ninety-nine times out of a hundred, she's right on. (There was *one* time when she picked the movie for my whole family to watch on vacation, and it was so fantastically bad that we were tempted to throw fruit at her. But that was the one outlier.)

When we don't curate for beauty and instead feed a desire for distraction or pick based on messages alone, we miss out on the essential human need to *feel the world deeply*. Focusing on messages isn't bad at all, it is just incomplete. It isn't enough to hear truth in words; you have to feel truth in your soul.

That's what curating for beauty does. It moves truth from the head to the heart.

Curating for justice. Curating for justice means looking for stories that tear us up over the way the world is broken and that make us fall in love with the way the world should be. When the Bible talks about justice, it's talking about something way, way bigger than what government should do about crime. That is minuscule compared to the vast vision of justice in the Bible. The Bible talks about a comprehensive way that the world *should* be. It uses the Hebrew world *shalom*, which connotes the idea of "everything in its right place."

As a lawyer, I think of justice as a fabric or a great woven tapestry. When the fabric of justice is torn, people fall through the holes—usually the most vulnerable people. Weaving the fabric of justice means mending the holes for the sake of the vulnerable. But there is more. Justice as a tapestry reminds us that justice is beautiful. The tapestry of justice properly woven is something worth hanging on the wall of the world. We need to see justice on display because it's jaw-dropping gorgeous. This is why Micah wrote that we should not simply do justice but that we should *love* kindness (Micah 6:8).

Nothing in the world cultivates our love (or apathy) for justice more than stories. Every good story tries to tell us what's wrong with the world. Where is the fabric torn? What is broken in shalom? Who's falling through the cracks, and what can be done?

Every good story also tells us what we should do about this brokenness. Who is the hero who can solve this? And who needs to be defeated in order to restore the fabric of shalom? What does the world look like when it is restored? These are all questions of justice.

So every story is trying to make us feel busted up about something and makes us fall in love with a solution. The problem is when they stir up fear over the wrong things or stir up love for broken solutions.

This is where we must pay close attention to media, particularly the stories of new media. The streaming stories of news media are financially incentivized to run short stories whose very nature is to stir us up and keep us coming back. It doesn't matter that the story leans liberal or conservative; both are assaults on justice when the point is to consume more media, rather than to point us to the world that needs mending.

One of the first ways to curate for justice is to devote little to no weekly media time to short-form news and instead read long-form sources. These will be sources that you likely have to go find or pay for, because the market doesn't push them to you as aggressively. These long-form articles, podcasts, and documentaries are more likely to inform us about our vulnerable neighbors, stir our hearts toward them, and send us out into the world instead of just boomeranging us into more media.

A second way to curate for justice is to pay close attention to diverse voices in the body of Christ. I often seek advice on what to watch or listen to from my African American brother-in-law, Dan, because I know his outlook on the world is different from mine in important ways that I need to hear. It's important to note that Dan is much more than a source of media recommendations. He is my brother and friend. Our lives are delightfully intertwined even if he never told me who I should follow on Twitter or what podcasts are a "must listen." But I also know that Dan, as my brother in Christ who is black, has developed an outlook on the world which I trust, so I'm eager to lean on him and many others to help me see the world in a more comprehensive and just way. We can't curate for justice when we look at stories through only our own eyes. We need the diversity of all of God's children and the collective wisdom of many voices. This is always true, but it is particularly urgent at a moment when the mediums of news are designed to push us into echo chambers. We need counter-formative habits of diversity to resist the slide into tribalism.

A third way to curate for justice is to look for stories or sources that you know will stir up your heart to love the vulnerable. For example, I love listening to baseball on the radio, which I count in my weekly media intake. (I see all sports as live stories of competition unfolding before our eyes, which is why we love them so much.) Listening to a game of baseball is an incredible way to relax and bond with friends, but it doesn't do much for my sense of loving justice. When I listen only to the Washington Nationals,

I'm much more inclined to browse sports sites and learn about bullpen solutions than I am to read about the opioid crisis in the neighborhoods surrounding me or to pay attention to the new bus routes up for vote in my city that will affect the way the vulnerable can or can't get jobs.

Stories can become a distraction from loving justice, because we would often rather be numb than soft. One of the ways I've countered that is by making sure my curated media feed includes things other than sports, like YouTube videos of Dr. Martin Luther King's speeches or TED talks by Bryan Stevenson, a trusted voice on the contemporary problems of criminal justice. These curated voices prick my heart, which would much rather be numb. They remind me that the point of limiting media is not *at all* to ignore the problems of the world but rather to open your eyes to enter into them.

I believe a new problem of my generation is the way that (whether right or left leaning) the ever-outraged and always-offended tone of mainstream news sources is making us numb to the world's pain. When everything is a crisis, nothing is. We think we're becoming informed, but actually we're becoming numb. True understanding of the world's brokenness and real compassion for the oppressed will not come from the firehose of online anger but from a careful curation of the love of real justice. We must resist becoming people who talk of justice out of rage and work on becoming people who talk of justice out of love.

This endless stream of media *will* drown out the quiet cries of the vulnerable unless we curate specifically to hear them, to love them, and at some point, to close our screens and walk out our doors to where they are.

Curating for community. Curating for community means realizing that fundamentally our stories should be pushing us out of isolation, not into it. The vast quantity of addictive media poses a real danger. It captures our hearts with really good stories but at the cost of spending our lives on the couch. This is not a new struggle—TV has been at it for decades, but it's newly intensified by the streaming story.

One of the baseline practices here is to watch most, if not all, media *with* someone else. This may be a roommate, a spouse, a neighbor, or a friend. Stories are an enormous part of our common bond. Sharing and loving the same canon of stories is one of the most significant ways we create social cultures. Community also filters what you watch—we are more discerning when we choose together.

We can lean into this reality by inviting our neighbors in to watch a TV series or by calling a gathering of friends to watch a movie that's particularly meaningful to you.

Everyone has different tastes, so curating for community means changing and potentially even lowering your standards to watch something you otherwise wouldn't, because it also means being with a person you otherwise wouldn't. Wisdom is needed here, of course, but I'm more skeptical of a wisdom that separates us than a wisdom that binds us, and I will always err on the side of community.

LOVING NEIGHBORS THROUGH MEDIA

When I think back to standing in front of that shelf in Daedalus Books, I realize that what I saw then is strikingly similar to what I see now on any computer screen: a wall of stories. The world is still made of stories, but my perspective has fundamentally changed. Then it was dread that I couldn't have them all. Now it's a desire to find the most beautiful ones and the most just ones, and to pursue them in ways that push me closer—not farther—from other people.

This is a desire cultivated by habit. I don't think of this as personal virtue; I think of it as a lens on the world, an act that has public consequences.

The habit of curating media is a habit of neighbor love because stories send us out into the world as certain kinds of people. Curating beautiful stories means we live in a slightly more beautiful world. Curating just stories means we collectively attune our sense of justice and turn our eyes to the vulnerable. Curating for community means resisting isolation and moving out toward our neighbors and friends.

Curating stories is not just about reallocating your time. It's also about reminding yourself that there is one true story. It's about retraining yourself to see that any good story will reflect the one true story in some fundamental way.

To curate media is to build a life that learns to cry, laugh, and applaud along with the gospel story that is tragedy, comedy, and fairy tale all at once.

WEEKLY HABIT 2
CURATE MEDIA TO FOUR HOURS

THE HABIT AT A GLANCE

Stories matter so much that we must handle them with the utmost care. Resisting the constant stream of addictive media with an hour limit means we are forced to curate what we watch. Curating stories means that we seek stories that uphold beauty, that teach us to love justice, and that turn us to community.

THREE WAYS TO START

A time audit. This can be an intimidating habit, so perhaps start with just a time audit. I find keeping track of my time is an amazing way to show me what I care about. Track your media watching for a week, and then set an hour goal for the next week that is four hours, or some other reasonable number you feel comfortable with.

Make great lists. My wife, Lauren, always keeps a running list of great movies, shows, or podcasts to listen to. This is a great way to begin the habit of curating. Instead of watching "what's on" or something Netflix suggests, have a list of things worth watching that you can go to and work through. If this seems hard for you, search for lists online from people you trust. There are really good ones out there.

Turn off auto-play. In a film class in college, our professor once energetically argued that credits are part of a film. You need the time, the music, the words, and more to come out of a story and reflect on it. Not only do auto-play functions ruin that reflection on a story, they also—as the programmers are aware—highly incentivize the choice to watch another story. Go to settings and turn them off.

THREE CONSIDERATIONS

Great stories. One of the best ways to curate stories well is to encounter great stories. They change us. They both satisfy us and make us hungry for more great stories. Often novels or movies that have stood the test of time are the best places to start. Notice also that the written form resists addiction because of the time you have to put into it. Balance your media by also reading a great novel or biography.

Read your feed. The stories we watch say a lot about us. They tell us who we are and what we love. Take an honest look at what YouTube predicts you want to watch in their recommended videos or what Instagram thinks you like to look at. Is that who you want to be and what you want to love? But don't judge yourself too harshly. This isn't to make you feel bad (and maybe you like what it says about you). This is the beginning of curation. Find out what you watch—Big Data already knows it better than you do—and then add, cut, and curate from there.

Video apps. I watch plenty of YouTube videos, but I find I choose more carefully when I don't have the app. Same goes for Netflix. I also watch on a computer, not a phone, because I'm resistant to the private feel of a phone. I choose better when I choose in public spaces. For that reason, I keep a program on my computer that shares my entire internet history with a couple other close friends. The point here is to figure out what helps you curate the best and to notice that usually we curate worse when we curate in private.

FOR BETTER OR WORSE, WE WILL BECOME THE STORIES WE GIVE OUR ATTENTION TO. IN A WORLD OF LIMITLESS STREAMING STORIES, WE MUST SET LIMITS THAT FORCE CURATION. THE WEEKLY HABIT OF CURATING MEDIA HELPS US CULTIVATE THE ABILITY TO CHOOSE STORIES WELL.

READING AND RESOURCES

Movies Are Prayers, Josh Larson

Telling the Truth: The Gospel as Tragedy, Comedy, and Fairy Tale, Frederick Buechner

FAST FROM SOMETHING FOR TWENTY-FOUR HOURS

To fast is to suffer.

DAVID WILLIAMSON,
MY FRIEND AND BROTHER-IN-LAW

"When you fast . . ."

JESUS (MATTHEW 6:16)

One spring, a bunch of our friends took a long weekend down in Jacksonville, Florida, to see our dear friends Matt and Kami. Matt is my best friend (the one you read about in the habit on friendship) who has been in the Navy there for years, and we've made a tradition of taking a trip down every other year in the spring. Some of us have kids, some of us don't, some of us are single—we all go. For a long weekend, we lounge around the pool, drink tiki drinks, and make slow extravagant dinners.

At one particular dinner that year—which included wine, homemade bread, and some smoked BBQ we had worked on all day—we sat outside on the deck, talking late into the night. We had come to the sweetest place of long friendships, where we slowly work our way around the table, catching each other up on our lives.

But as often happens in these most beautiful times of friendship, sadness was close by. Two of our dear friends—for whom we had been praying for months—had just found out that day that once again they were not pregnant. Now, for almost a year, they had gone up and down in the cycle of hope and prayer. But once again there was nothing to report but sadness.

Then, as we kept sharing around the table, another two friends talked about grieving their miscarriage, which had happened just a few months prior, and their struggle to conceive since.

The tears and the pain stood out with a particular contrast before the feast we had set. We had strung lights and made cocktails, we had bought and cooked everything we needed to fill our stomachs, and yet the nagging emptiness of the heart persisted.

There will always be a kind of pain that you can't soothe by eating.

FEASTING AND FASTING

There was food in the world before there was emptiness in the world, and that's an important fact. When God made the Garden of Eden for Adam and Eve, he didn't put the fruit there and tell them to make sure they got three square meals a day or else these bodies that he made for them wouldn't work. No, he set them in the garden, where "every tree . . . is pleasant to the sight and good for food" (Genesis 2:9). God's people had no lack, and they had food because—well—God is generous, and he makes wonderful things.

We were made to feast. Not in order to become full, but because we *are* full. We are to celebrate that fullness by feasting. Feasting to fill the emptiness is not feasting; it is coping.

But the fall changed everything, of course, eating included. And the best way to understand the fall is to say that we take the good things God has given us and make them god instead. When Adam and Eve ate the fruit, they inverted God's gift. They ate to become God instead of to celebrate God.

Among all the other death and suffering that entered the world after the fall, our relationship with food was also broken.

Now we have to toil with the ground to get food. Scarcity and fruitless labor have come into the world. We have bodies that will die unless they're fed, and the first murder came because of jealousy over food.

What was meant to be the culmination of the celebration of life with God became the mark of our inevitable suffering and death.

EATING AWAY OUR EMPTINESS

In a world of suffering and death, one of our greatest temptations is to rehearse the fall again and again through food. We eat to try to fill our emptiness. This is why fasting is mentioned so often in the Bible. Fasting is a way to resist the original sin of trying to eat our way to happiness and to force ourselves to look to God for our fullness. In that sense, to fast is to lean into the truth of the world: we are empty without God. "Man does not live by bread alone" (Deuteronomy 8:3).

In fasting, what begins with experiencing the emptiness of our stomach ends in experiencing the emptiness of the world. In the Bible, fasting is not just to reveal and clarify our own need for God. It is to lean into the suffering of the world itself and to long for God to redeem it. This is why the Israelites during the time of Esther fasted; they knew the brokenness and injustice of the throne they lived under, and they longed for God to redeem it.

This is partly why Jesus fasted before he began his ministry. He was sent to undo the fall, and his forty-day fast was an act of longing for the world to be restored by his ministry to come. He emptied himself in a bodily prayer for the world's fullness.

In this way, fasting tracks the plotline of all things. We must be emptied in order to be filled. Christ denied his body so we could partake of his body.

The weekly habit of fasting, then, is a way to lean into both the emptiness of the world as it is and prayer for the coming fullness of the world as it will be. The world doesn't end in fasting, of course, but in a feast. Above all, we fast because we long for the wedding supper of the Lamb.

HOW FASTING REVEALS OUR INNER NEED

After that night with friends in Jacksonville, we decided to fast together. One goal of this fast was to share in the emptiness of our friends who had empty wombs, but another part was to pray and long together that the world and their wombs would be filled. Over an email chain, we picked a day. Everyone did it differently. Some of our friends were pregnant, nursing, or otherwise unable to fast from food, so they chose to fast for twenty-four hours from sugar or social media—anything that would make them feel the longing. Some of us have bodies or jobs that would only practically allow fasting from one meal. Others of us went for twenty-four hours or even more without food.

The point of fasting is not the technicalities of it. The point is leaning into the lack, and this can be done in many, many ways, all of which are radical acts, especially in America. Here, fasting is bizarrely countercultural because it runs the opposite direction of the American dream. In pursuit of the dream, we tell each other that we can move upward in the world through sheer individual effort and that we're going to finally be happy when we get there. In fasting, we deliberately move downward into emptiness—and even more, we admit that we can't eat or work our way to happiness. We need God for that.

It's almost trite to talk about the American way of comfort because, well, it's so obvious. We are so well fed that we're actually dying from our excess. We eat to avoid our problems, and then we eat to deal with the problems that our overeating caused in the first place.

Fasting exposes all of this. It suddenly exposes the self because you can't use food to dull your desires, numb your feelings, or make you feel satisfied or happy.

My emptiness is the first thing I notice when I fast. No matter how regularly I do it, it always hits me like a surprise. "Oh no! I can't eat today." And I immediately feel just plain depressed. On any other day, I don't think about the way breakfast sets my emotional tone for the day. During a fast, when I don't have breakfast, I honestly think, *Well, what's this day worth anyway?* And the hunger hasn't even set in yet.

It's midmorning when I become irritable. Not only am I trying to concentrate over a hungry stomach, but I also can't do what I otherwise do every day: look forward to lunch or snacks as a way to medicate the pain of toilsome work. So then I move from irritation to anger, and impatience leaks out of me everywhere. And that's the first point of the practice of fasting: seeing who you really are.

When I fast I see that deep down I'm not actually a very patient person after all. I'm not actually a very content person after all. I'm not as independent and strong as I thought I was. I'm a weak, impatient, angry person who medicates with food and drink. This is painful to confront. Yet to live without fasting is to live without knowing who I truly am.

And this is not the end of the story, for man does not live by bread alone, but "by every word that comes from the mouth of God" (Matthew 4:4). That is to say, there is true life in true emptiness.

When I fast, I try to set aside some time at each mealtime to pray. The prayer begins to fill me in a way that, while my stomach still hurts, my soul is filled in an unusual way. Fasting is to let your desires hang out in the open, where you can observe them. In combination with prayer, this often leads me to a kind of spiritual clarity that is impossible without the act of fasting. I see more who I really am, and I feel more of who God really is.

Slowly but surely, my posture in relation to the world is recalibrated. I'm not here to get what I want; I'm here to love other people.

Often when I come home in the evenings, I am consumed with what our dinner plans are. But when I'm fasting, a monumental switch happens. I come home not expecting to eat. I'm simply expecting to serve other people as they eat. The most remarkable part is that I'm actually happier that way, because all along I've been thinking that food makes me happy. But now I see that only love does that. When that switch happens, ironically one of my favorite things to do is cook for other people while I'm fasting.

HOW FASTING REVEALS THE WORLD'S NEED

On one of the nights that I was fasting with my friends after our Jacksonville gathering, I decided to take a walk. Evenings are difficult when I'm fasting, as the day dies down I'm left with the bare hunger. Often I can't figure out what to do with those moments, and because I usually try not to work on the evenings that I'm fasting, I'm left with nothing to distract me. In order to do something about that, I will often listen to something or maybe take a walk and pray. On this particular night, I did both.

At the time I was living in a neighborhood on the north side of downtown Richmond called Jackson Ward. It was close to my law firm; a ten-minute walk or five-minute bus ride south would get me to the office. But as I took a walk that evening, I realized that every time I walked out of my door, I walked south. I did that because that's where my office is. That's where the YMCA where my boys swim is. That's where the public library is, where we read and play. That's where the bars are my friends and I have drinks at. That's where the good coffee shops and bakeries are. That's where the bus stops are.

In short, the life of the city is south of my house. And it's so vibrant that I never gave much thought to the fact that I never walked north—until the night when I did.

I walked north because I had just finished listening to Dr. King's address at Stanford in 1967, in which he described two Americas: "Every city in our country has this kind of dualism," he said, "this schizophrenia, split at so many parts, and so every city ends up being two cities rather than one. There are two Americas."[1]

As I listened that night, I knew what he was talking about because I lived on the border of the second America. So that night I turned north instead of south, and I began walking the three blocks toward Gilpin Court, the largest public housing project in Richmond. It's closer to my house than my office. I walk to my office every day, but I had never walked to Gilpin Court. Not once.

The life expectancy in Gilpin Court is twenty years less than in the neighborhood two miles southwest of here, and roughly one of every three males born in Gilpin Court will go to jail. The internet does a nice job of summarizing the impression the words "Gilpin Court" gives. A google search of just those two words at the time of this writing brings up the results of "massacre," "stabbing," "fatally shot," and "gunned down in random daytime shooting" all on the first few hits. As far as a word count on the browser goes—and I actually counted—there are far fewer mentions of *death* in a search for Afghanistan or Syria.

Richmonders say that my neighborhood, Jackson Ward, used to be one of the most thriving African American communities south of Harlem. Some called it the black Wall Street. In the mid-1950s, it was proposed that an interstate be built right through the middle of it. It was voted down by popular referendum because of the number of African American residences that would have to be destroyed. Then the state General Assembly, removed from local concerns, pushed it through.[2] The center of Richmond's African American cultural and commercial life became a war zone of poverty and projects. In the forties and fifties, public housing was built to try to clear out slum housing. Of course, we know the end to that story. The combination of public housing concentrating the poverty and of the interstate isolating residents from the economically thriving part of the city created a Third-World-type nightmare within our Pleasantville of Richmond.

How we got there is terribly complicated, but the reality of living in the area is terribly obvious. We've busted the city up. This side of the bridge is safe. That side is a war zone.

On that night, as I walked down the two blocks and stood at the bridge that goes over the interstate, I kept thinking about King's two Americas. I could see into that America. A man in a hoodie passed me. I watched his movements carefully. He watched mine. Neither of us could see each other's eyes. Across the bridge are vacant lots and no trees. There's a blinking sign advertising malt liquor at the only open store in the neighborhood. Through the bars on its window, I can see some men milling around. Around the corner I see a flashing blue light and hear police sirens.

Dr. King gave his "The Other America" speech more than fifty years ago, trying to call attention to the radically different lives and opportunities Americans have. Yet I live a life of prosperity on this side of the interstate and that night, my feet stopped at the bridge. Try as I may, I couldn't bring myself to walk across.

Suddenly overwhelmed by the weight of all that we've done to each other, I sat down on a bench a block away, and began to pray. "Lord," I remember saying, "tell me what to do. I know that you know, and you can tell me. Just tell me what I can do about this. Right now, please speak. I'm listening. Just say something."

Then I waited.

I was a few days into this particular fast, and I felt an intimacy with the Lord that made me certain he was going to speak. I sat in stillness, leaning forward. "I will tell everyone," I prompted, "just tell me now what we can do."

Suddenly I sensed light all around me. The mood of my quiet changed, and my heart did a small leap. I could feel something happening. When I looked up, I saw a light. It was red, white, and blue, and there was a rush of noise. An ambulance and a police car whipped around the corner, sirens blaring. I could feel the rumble of their engines. They accelerated and sped by me, lights spinning into the night of Gilpin Court. Then I lost sight of them, and all was quiet again.

That's it? I thought. I stayed quiet for another long, sad period of time. Then, with nothing but silence and sirens, I walked the two blocks home.

The next Saturday, I was driving by the same spot with my wife and sons in the car—to the interstate, of course. The light turned red and I stopped. Just outside the window was the bench I'd sat on while waiting for an answer but heard only silence.

As I was reimagining the disappointment of that night, I began to hear my name being called. It was being repeated, over and over, softly but insistently. "Papa? Papa? Papa!" I snapped out of it and realized my son was calling me. He wanted a book off the floor. I reached back and handed it to him. He fell quiet and began turning pages. I looked at him, barely two years old and flipping through books, well on his way to learning to love to read. It suddenly dawned on me how this two-year-old was in a real way more powerful than most anybody on that other side of the bridge. Why? Because he had the power of being heard. When he calls, someone answers.

It's clear to me now why God answered me with silence that night. Silence is the hallmark of the vulnerable. They are vulnerable for many reasons, but this may be the main one: When they call, no one answers. And it's not because they can't talk, not because they don't have something important to say. It's because we have slanted the system to drown them out. The tapestry of justice is torn in just the right way to obscure their voices.

You can understand why they stop calling. They don't believe in the power of it anymore. The vulnerable are the vulnerable precisely because they sit in silence and have no voice to reach the deaf ear of power.

We could talk about how everyone is vulnerable, certainly. And it would be true. We all are in real ways. But saying that flattens the reality of the things; it papers over the truth of the world with metaphor. Metaphor should be used to point at truth, to draw it out, not to obscure it. So that wouldn't be right here.

The fact is that there is a particular sort of "it doesn't have to be this way" suffering that goes on in dehumanizing material poverty. It's the way it is because *we have made it that way.*

It's good to let that hang out in the open and name it. And fasting allows us to look with spiritual clarity on who we are and what we've done, to name the realities, no matter how uncomfortable it is.

When we fast, we become more attuned to the stubborn reality of the world's suffering. Many of our neighbors' sufferings are hanging out there in the open, and they need to be seen. Confronted. Named. Especially by those like me who *do not* suffer as they do.

Even when we don't know what to do or how, it's important simply to cut through the silence. That's a step toward loving justice.

So while there is a part of fasting that reveals our own need, there is a part that reveals the world's need too. To fast is to lean into the emptiness that seems to run under all of us and simply to say, "Yes, it's there, and it's there for my neighbors too." Fasting is a way to quiet the noise of the world and listen for the "rumble of panic underneath everything," as anthropologist Ernest Becker put it.[3]

This kind of fasting is a way to lean past our own emptiness and into someone else's. It's a practice of empathy, of willingly walking into pain for someone else. It's an imitation of Christ, limiting ourselves for the sake of someone else.

HOW FASTING REVEALS THE ONE WHO MEETS ALL NEEDS

Their names are Juniper and Nathaniel. At the time of this writing, they are both one and a half years old, and they are beautiful. They were conceived into the world in the next two months after Jacksonville.

As it went, we fasted once, and at the end of that month, the couple who had miscarried became pregnant. We rejoiced. We couldn't believe it. But it still wasn't enough. So we fasted the next month. That month the couple who had never been able to conceive became pregnant. We were amazed. We rejoiced again. We feasted.

Life is usually not so tidy, but these babies, Juniper and Nathaniel, are both little Ebenezers. They are signs of what the Lord has done, and these two answers to our fasting and praying are worth sharing. It was so wonderful that we decided we should fast a third time together. We said, "Our prayers have been answered, so let us fast for the world's need."

That was during the time I took a walk to the edge of Gilpin Court. And here is the ragged edge of the story: Nothing changed. Not that I can see.

Yes, I listen differently now. Yes, I have been changed. But I must not slant this the wrong way.

There is still so much more I want to see redeemed in my small corner of the world in Richmond. And these problems of poverty and vulnerability seem as immovable as the James River that runs through my city.

The reality of this side of fasting is that many of our deepest prayers aren't answered with miracles we long for. There are two healthy children, but the poverty of my neighbors lingers. I have a renewed sense of what it means to listen for the vulnerable, but the sirens still sing their sad song night after night.

CHRIST IN THE EMPTINESS

Even when the prayers aren't answered, in the space that fasting opens up, we find something else: Christ. He is the one who can and will fix all of this, and the one we are to be *with* while we make an effort to do the same.

In emptying ourselves, we practice becoming like Christ, who emptied himself. We practice sharing in his sufferings. It isn't on the mountains of triumph and victory but in the valleys of sorrow and loss that he waits for us.

And this is illuminating to the Christian faith in general. It is possible to be a people who avoid our exile within America and constantly lean into assimilation into our culture. We can easily forget that the place of friction, persecution, minority suffering, wilderness, emptiness, and hunger—*that is the place where Jesus is.*

Fasting, as a practice, is a way to enter into Jesus' life. He was a homeless, hungry minority. He was a refugee and an outcast. He was actually—not just metaphorically—poor. He lived among violence. He died violently. To follow Jesus is not just to believe in his life; it is also to follow him into his lifestyle. And that idea runs hard against my usual expectations of being American. It's hard to be poor in a land of plenty, and it's hard to be hungry in a land of sugar. It's hard to be empathetic in a land where we hide the poor on the other side of the interstate. It's hard. It's all just hard.

But fasting is a habit of breaking that comfort in order to seek true comfort instead. We continue to fast because that's where we find Jesus—right on the fault line of the beautiful and the broken. The miracles we see in fasting are amazing. And the brokenness we enter into in fasting is unbearable. This is, I suppose, why the practice of fasting is a beautiful and painful reminder of a good world cracked by the fall.

Cultivating the habit of fasting as a way of life means cultivating an understanding of why beauty and brokenness intertwine in the present world. I know of no other way of life that can both acknowledge all that the Lord has done and still yearn for all that we desperately long for him to do.

WEEKLY HABIT 3

FAST FROM SOMETHING FOR TWENTY-FOUR HOURS

THE HABIT AT A GLANCE

We constantly seek to fill our emptiness with food and other comforts. We ignore our soul and our neighbor's need by medicating with food and drink. Regular fasting exposes who we really are, reminds us how broken the world is, and draws our eyes to how Jesus is redeeming all things.

THREE WAYS TO START

Pick something to fast from. The first step is choosing something that's helpful to fast from. For regular fasting, I prefer simply fasting from all food. But fasting from sugar, meat, alcohol, caffeine, social media, TV, internet, or something else may be a good way for you to begin regular fasting.

Sundown to sundown. My preferred fast begins at sundown on a Thursday and ends with a communal breaking of the fast at sundown on a Friday. This is a great way to do a twenty-four-hour communal fast with friends.

Start with a meal. If fasting intimidates you, a great place to start is just skipping a meal—maybe lunch—and replacing it with prayer. If you do it with family, skip your family meal so you can pray together. If you do it with friends at work, choose lunch. Being in this rhythm with someone else makes it easier to try fasting, and the communal nature of it changes the experience.

THREE CONSIDERATIONS

Communal fasting. I find fasting far richer in community—not to mention it's so hard to muster the discipline when it's just me doing it.

Consider having a text or email chain going among people who are fasting so you can share encouragements and prayers. Also consider doing an initial prayer time together and/or breaking the fast together.

Prayer. Usually, just skipping meals doesn't lead to prayer. I find that I need to take walks instead of eating to actually pray. Whatever you do, make sure you replace meals with prayer.

Multiple days. This is nothing to rush into, but I've had wonderful times meeting the Lord during fasts for multiple days. I've also had difficult times where fasting for multiple days was the way to repentance. There's something unique about the state the body and soul enters into on longer fasts, and I would recommend working toward them as an occasional practice.

> TO LIVE WITHOUT FASTING IS TO LIVE WITHOUT KNOWING WHO YOU TRULY ARE.

READING AND RESOURCES

Celebration of Discipline, Richard Foster

A Hunger for God: Desiring God through Fasting and Prayer, John Piper

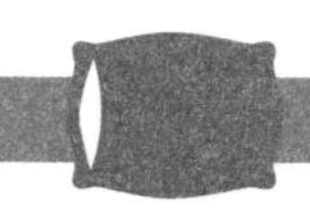

WEEKLY HABIT 4

SABBATH

A man who works with his mind should sabbath with his hands, and a man who works with his hands should sabbath with his mind.
ABRAHAM HESCHEL

The sun will rise
To your surprise
All by itself
Without your help.
THE HILL AND WOOD, "ALL'S WELL"

I stood in a small room while a Chinese doctor mumbled and made notes on a clipboard I couldn't see. He was wearing a white lab coat, which I found slightly humorous and comforting at the same time.

"How much you sleep?" he asked me in his broken English.

I hesitated, debating whether to tell him about the spreadsheet I was using to keep track of my hours of sleep each night. But since the point was to minimize sleep and make sure I kept it to five hours or under, I didn't bring it up.

"Probably not enough," I responded in Chinese.

"You work lot?" he asked. I hesitated again. I was frustrated that he was speaking English. The reason I was barely sleeping and working around the clock was that I was trying so hard to learn Chinese.

The great historical missionaries to China were famous for how well they learned Chinese. It was said that Matteo Ricci, the sixteenth-century Italian Jesuit missionary to China, would ask anyone at the emperor's table to recite an ancient Chinese poem. After hearing it once, he could recite it back to his listeners verbatim—except backward. Ricci was a master of memory, and his diligence gained him incredible favor and opportunity in China to teach the highest courts in the land about Jesus.

In China—and in many other places in the world—learning the language of a people is a way to show honor and respect; it's the way to speak a language the heart understands. Once I called an air-conditioning repair company in China to request a repair. When they came to the door, they said they had the wrong address because the person who'd called them was Chinese. It was the proudest moment of my stay there.

Legend has it that Ricci had what he called a memory palace—a vast castle he pictured in his mind where he made a mental map of all the things he learned. When he heard something new, he pictured setting the facts in a room in his palace, and when he needed it, he visualized walking back in the room and picking up the knowledge. I was then, and am still now, completely enamored with this vision of the endless mind.

I desperately desired to be a good missionary, so I dedicated myself to learning. In service to that I began trying all kinds of strange life hacks, trying to figure out ways—in my own efforts to be like Ricci—to push my body and mind to do more.

I tried intermittent sleeping and other methods of minimizing sleep. I bought books on speed reading and tried to train myself to read a page in a matter of seconds. I also began keeping track of my time in fifteen-minute intervals, so I could see where it was being used. I was scheduling every morning for reading, news, writing, and study so that by the time I left for class at eight thirty, I had already done more than most do in a day. I even tried creating my own memory palace.

But it was mostly useless.

I kept crashing when I tried not to sleep, and every time I read a page in thirty seconds, I came away not having the faintest idea of what it said. Sometimes I didn't even know what book I was reading. I walked my palace over and over, like a ghost haunting its hallways, but I couldn't even remember the rooms, much less what was in them.

Despite the disappointment of hitting the limits of my mind and memory, there was still a real pleasure to this period in my life. Keeping track of my time was working. It was like budgeting money—once I saw where the asset was going, I started acting differently. It was helping me learn Chinese faster. I still keep track of my time (which I realize is weird).

There are few deeper satisfactions than throwing yourself headlong into good work. In fact, the paradox of good work seems to be this: Anything worth doing requires bending your whole life toward it. On the other hand, nothing is worth bending your life until it breaks. I never seem to know where that point is until after I break, and that season in China, I broke. That's why I was in the doctor's office.

Late in the spring of that year, I began to develop persistent lesions—areas of swelling and inflammation that seemed to have no cause but wouldn't go away. When one finally developed right over my cheekbone, I decided to go to the doctor.

"Yes, I'm working a lot," I said. Then I cracked. I told him about my schedule, about the sleep minimization, about the drive that was putting my body under so much stress.

He put down his clipboard and for the first time in the conversation switched to Chinese. "*Bu yao ji,*" he said. That's a casual way of saying, "Don't worry."

He put his hand on my shoulder. As he spoke in his own language, his personality suddenly shifted. He sounded tender and wise. He seemed less of a doctor and more of a grandfather—someone I could trust.

"*Ni xuyao xiuxi* (You need to rest)," he said with a smile.

An incredible sense of relief spread over me. And then I went home.

A GOD WHO RESTS

In the beginning, God created, he pronounced it good, and then evening came. He did this six times before sitting back. On the seventh day, he stopped. He did nothing. He sabbathed because the work was finished.

Amazingly, most people in the world today still organize their lives around this divine seven-day rhythm. We work, and then we call it done for the weekend. There's a reason for this. We were made for this rhythm.

Focusing and finishing are the two great glories of work. My best days at work are either when I pour myself into something and become lost in the pleasure of focus or when I finally complete a project and get to check it off my list. My wife and I often joke (or more accurately, complain) that the hardest part about spending a day parenting young children is that you never get to do either.

But the rhythms of focusing and finishing seem to be built into the DNA of what it means to be a human being. This is exactly why treating our bodies like machines is wrong; they weren't made to work without consistent and rhythmic pause points when we finish and rest.

When I look back on my phase of life hacks, I see something sinister in the very language of it. The fundamental idea was that I was a computer or a machine, and I could find ways to work around my inadequacies. I could find solutions for my pesky limitations, like the need for sleep or rest.

My envy of Matteo Ricci has always been strange, so I don't blame him. But what I loved about him was the way he seemed to be limitless, which is to say the way he was unnatural. I love this because I don't like my limits.

I admit I'm odd. I have always been prone to eccentricities. The speed reading and the sleep tracking are things I don't usually tell people about for a reason. But the impulse that led me to those things seems to be universal. None of us like our limits. Like Adam and Eve in the garden, we are not content to be *like* God; we want to *be* God. The weekly habit of sabbath is to remind us that God is God and we are not.

BUSYNESS AS A STATUS SYMBOL

Our culture still arranges its time around the seven-day schedule, but what is now conspicuously missing is resting on the seventh day. In fact, in professions and careers like mine—and there are many—the very idea of taking a day off is at best quaint and at worst scandalous.

Once I was giving a talk to a conference room of sixty or seventy new, young lawyers at my firm. When I told them that in order to be good lawyers in the long run, they needed to cultivate a habit of taking one day off every week, you could have heard a pin drop. It was as if I was speaking of scandal. They were intrigued, maybe even hoping they could hear more, but at the same time wondering if a senior partner of the firm, who was standing next to me, was going to yank me off the stage.

I got the sense that if I had been more like my China self and told them about new ways to life hack the need for sleep away, they would have nodded in solemn approval while dying just a little bit more inside.

It's like the way I looked at Ricci. We seem to have come to a point as a culture where we praise the acts of being inhuman as acts of being a great human. The consequences, of course, are dreadful.

That summer I had to get minor surgery to remove the lesion on my cheekbone. I still bear that scar of unrest. But I'm very grateful for that failure because that's when I learned that rest is a generous gift. It's made for us, for our body and our soul. When I tried to live outside its limitations, my body and soul both suffered for it.

Some people aren't so lucky. They spend great chunks of their life denying the need for rest until they crash, often much more spectacularly. I see this happening in my workplace daily, but I see it happen in my church too.

We may take weekends, but our days away from work are often spent furiously trying to accomplish other things: new hobbies, new travel, more hangouts, more side hustles. We have a spare day, so we need to "get our life together" or do all the things around the house we've been meaning to take care of. Stopping and taking a nap would be a sign of weakness or poor stewardship. Sometimes we honestly feel it's immoral to rest.

It used to be that the upper class showed off their status by displaying their lives of leisure. Now we do it by conspicuously displaying lives of constant busyness. The more important you are, the more in demand your time is, so nobody who is anybody has time for enough sleep.

And this, of course, is what we in our restless culture are after: an abiding sense that someone thinks we're important.

THE RESTLESS SOUL

There is more going on than just our body's need for rest. Our souls need rest too. But the rest that our souls need is not simply a nap. It's the rest that comes with realizing we don't have anything to prove anymore. We don't have to prove we're important.

This is why we live in a culture that can't accept sabbath; we do not believe that work is *from* God and *for* our neighbor. Instead we believe that work is from us and for us. It's something we pursue to become who we want to become. Our careers define us. This is the American dream. We can work our way to significance. This is what we're doing when we prove our busyness to ourselves and each other; we're trying to show that we matter, that the world wants us, that *the world depends on us*.

But the gospel wants to put that to rest. We don't have to work like that because Jesus has done that work for us. And he has finished it. The book of Hebrews tells us that God has entered into his eternal rest, which is another way of saying that God has entered his full sabbath, because his work is done. This is because God has not only finished the work of creation, now in Jesus he has finished the work of redemption too.

OUR SOULS NEED MORE THAN TO DO NOTHING; THEY NEED TO DO *RESTFUL* THINGS. IN THIS SENSE, REAL REST TAKES REAL WORK.

When the project went wrong because of our sin, God came in to fix it in the form of a baby. When Jesus came to live and die among us, he came to finish the work once and for all. This is why, in his final words on the cross, Jesus cries out, "It is finished!"

What is finished? The work of salvation. In his death and resurrection, Jesus has done everything needed to unite us back to the God who loves us. There is not a single thing to add. But there is everything to receive.

The rest beneath the rest is the knowledge that in Jesus all work is finished.[1] This is why Augustine wrote, "My soul is restless until it finds rest in thee."

When that's true, we can finally take a day off. We can finally take a nap or stare at a cloud or have a long dinner with friends. "It is finished" is the lullaby of all things, our restless hearts included.

THE SABBATH: ALL THE DOING IN NOT DOING

As I learned in China, everyone has to rest. If you don't choose to rest, then reality will make you rest—often in the form of sickness, injury, or emotional breakdown. These ways tend to be painful. After that year in China, Lauren and I began to practice sabbath as a matter of necessity.

One of the first things we learned was that proper sabbathing is much more about doing than not doing. It's about doing restful things. Often our inclination to stop and veg out ended up being counterproductive. Some of it was exactly what we need, especially what our bodies need, which is some downtime and the refreshment of laughter. But our souls need more than to do nothing; they need to do *restful* things. In this sense, real rest takes real work.

In China, Lauren and I found that there were certain things that rejuvenated us from the week of language learning and conversations with students, and most of it took careful planning. An ideal sabbath looked like this: sleep in, worship, long lunch with friends, go home and rest, maybe nap, maybe make love, go out and explore some part of the city we hadn't been to yet or take a walk in a park, and bring a book that is pure pleasure reading.

What all these things had in common was not that they involve "not doing" but rather that they involved doing worshipful or engaging activities. They were things that drew us closer to God and others. The rest I needed was not only more sleep, but it was also the rest that comes with unfolding in good friendships or sitting still in God's creation.

Vocation and sabbath. Our habit of sabbath has changed over time and vocational seasons, and it should. Abraham Heschel, a rabbi who lived during the civil rights movement, put it like this: "A man who works with his mind should sabbath with his hands.

A man who works with his hands should sabbath with his mind."[2] What your work looks like is going to affect what your sabbath looks like.

For a long time, writing was a part of my sabbath. It engages my creative side and has always been one of the ways I relate best to God. In writing, I feel the pleasure of creating and changing through the power of words. However, when I began writing this book, I immediately realized that this was really, *really* hard work and that I would need to sabbath regularly from it. This caused a couple of failed sabbaths because I tried to fit in writing then realized that something that used to bring me rest was now ruining my rest. Sabbath changes as life changes.

When Lauren and I became parents, there were totally new challenges to our sabbath. If it's not obvious, you can't just take a break from children. Consequently we've realized two things.

First, there are seasons of sabbath. There are seasons when a sick parent, a newborn, a tough new job, or something else will make sabbathing really hard. But remember, one of the most important things to be done in the pursuit of habit is to focus more on the rule than on the exceptions. Developing the background rhythm of sabbath is the foundation. That means that the tough times—where we get out of our routines—become the unusual times, not the norm. However, it's really important to pursue sabbath precisely in those tough times, for those are the times we're most likely to run ourselves ragged.

Second, communal sabbaths change everything. Community can help you bear burdens in tough seasons so that you can sabbath even though a human life depends on you (for example, new moms). When Lauren and I didn't have kids, communal sabbaths often meant having a big meal with friends and lingering long to talk. At this stage in our life, doing a weekly Sunday supper with family means that while there are still roughly one million children to feed and take care of, we get to be together and help each other. Older kids can go off and play; younger ones can go off and nap. It's the most crazy and chaotic rest I've ever experienced, but it is undeniably restful. My mom deserves a mention here, for sabbaths are made infinitely more restful by her tendency to set out and clean up everything before the rest of us realize it's happening.

Also, with all the parents in one place, the ones who primarily spend all week parenting can take a back seat and maybe catch up on emails or something else that's restful—at least for them in that stage. Perhaps we could rephrase Heschel's quote for the contemporary household: someone whose work is parenting may sabbath with email writing. Someone whose work is email writing may sabbath with parenting.

Scheduling sabbaths. Sabbath, like anything else, comes with practice. And starting this practice when it isn't the norm can be terribly difficult. The first step is to pick a day and communicate it to the people around you. I found it very hard in certain seasons of law school and lawyering to communicate to fellow classmates or lawyers that I was going to wait until Sunday night to work. Sometimes it wasn't necessary to tell them, but once in a while I needed to communicate because of expectations. I've had to tell people that I wouldn't be responding to emails until Sunday evening. This has *always* been respectfully received, and even when awkward, the result is always better than not telling them at all.

Now Lauren and I keep sundown Saturday to sundown Sunday as our sabbath. Saturday afternoon is often filled with laundry, house cleaning, and all the other things we need to get done ahead of time in order to create space in the house (and the mind) for rest. Then, sometime in late afternoon, we inaugurate the beginning of our sabbath by lighting a candle with our sons. Marking time matters, and this practice has not only been a way to say, "Okay, we're starting now," it's also a wonderful way to get kids involved. (As it turns out, the boys really like fire.) Right after lighting the candle (and having fought about who gets to put the fire out and then having finally put the fire out), we go out and do something fun as a family. Saturday evening is usually spent in conversation with friends.

One of my favorite things to do Sunday morning is to get up and cook a big breakfast before church for the boys while Lauren sleeps in. This simple act alone makes Sunday special. We usually go to the latest service possible in order to preserve the unhurried character of the morning.

We also like the rhythm of Saturday evening to Sunday evening sabbaths because both of us have time Sunday evening to get ready for the week. I can always tell my coworkers that whatever comes up over the weekend that needs to be ready for Monday, I can work on Sunday evening after the kids are asleep.

DISCOVERING THE RESTLESS SOUL

Trying to sabbath brings almost everyone to the same realization: "I can't get it all done." Maybe it's the laundry, maybe it's the yard project, maybe it's work emails, or a job search. Whatever it is, when you plan to stop the work for twenty-four hours, you come to the stubborn reminder that you can't do it all.

> **PRACTICING SABBATH IS *SUPPOSED* TO MAKE US FEEL LIKE WE CAN'T GET IT ALL DONE BECAUSE THAT IS THE WAY REALITY IS.**

This is the point!

Practicing sabbath *is supposed to* make us feel like we can't get it all done because that is the way reality is. We can't do it all.

Sabbath protects us from acting out the lie that we can. Sabbath helps us discover the restless soul of which Augustine wrote.

I began sabbathing out of necessity; my health depended on it. Now I sabbath as a way to understand my salvation; my soul depends on it. When I stop working, I have to admit that the world doesn't depend on me. At times it feels like the planets will fall out of orbit if I cease to write emails and remove myself from the internet. Amazingly, not only do the planets faithfully hang in space but usually no one notices I'm gone!

Sabbath helps me see how small I am. When I don't see that, I'm always prone to misunderstand the reality of who is dependent on who. The belief that we sustain the world and God doesn't is at the core of our unrest. The violence of that belief shows up as scars on the heart and the body. I still bear mine on my left cheekbone.

In the deep stillness of habitual sabbath, the truth of the world begins to sink in: you are not necessary. That's the beauty of grace. In sabbath, we realize that even the most important of us can disappear, and the world will go on. But if Christ, the sustainer of the universe, were to disappear, everything would fade away.

Sometimes I think back to Matteo Ricci's memory palace. I picture how even the most brilliant, rich, and successful of us are still utterly dependent on the goodness of Christ to sustain the world moment to moment to moment. If Christ were to stop, then not only would the planets fall and all the light fade, but the sweetness of all memory, all the rooms in the palace, all the visions of our past and future—everything we hold dear—would pop and fizzle like a cut TV signal.

But Christ is generous. All things and the memory of all things live on. He sustains us still in this moment. And in this next one.

Sabbath, then, is the essence of our salvation. We can rest because in the end, as Julian of Norwich put it, "all shall be well, and all manner of things shall be well."[3] We can rest because God has done all that needs to be done.

COME AND SABBATH

I delight in the irony that it took a Chinese doctor to teach this American missionary to sabbath. When I remember the moment he put his hand on my shoulder, looked me in the eye, and told me in his own language that I was allowed to rest, I think of the person of Jesus. We are all looking for someone we trust to look us in the eye and tell us we've done enough, that it's okay to stop. This is the good news of the gospel.

Many, many people believe that being a Christian means trying to be a good person. The idea is that God supposedly likes "good" people. No one is *perfect*, but God at least forgives those who try hard enough.

This is not true! Don't believe a word of it. Even worse, it is the most burdensome lie ever told. Here's the truth: we are messed up beyond belief, but loved beyond belief, and that is the *one* thing worthy of our belief.

If you've lived your life believing that you can earn your worth, that you can earn your salvation by outweighing the bad with the good, that you can justify your place in this world through the money you earn or status you achieve—come and rest! Come and sabbath with Jesus. Here there is peace that no amount of effort can buy: he came to you first. He lived the good life we are all trying to live. He did it all. He sacrificed everything. He always said the right thing. He always knew what to do and where to go.

And where did it get him? It got him killed. People hated him. They stripped him naked and killed him. He lived the light of life we're all trying to live, and he was answered with death.

But it was all for love. It was all for you!

He stayed up all night in the garden of Gethsemane so *you* could sleep. He finished his work on the cross so *you* could rest. He let the world break him *so it doesn't have to break you*. He rose from the grave so all *your* aspirations won't end in the grave.

If you've read any of this book thinking you can muster the good life out of a few daily and weekly practices, you're reading it backward. Love has first come to us. Anything and everything else we do comes after. All these things are simply a response to this astounding love.

Christians do things, sure. Now that Christ is risen, there are all kinds of beautiful ways to live. There are all kinds of people to show his love to. There is much to study, and there are many languages to learn. There is so much good work to do alongside the Redeemer of the world. There are many habits to practice and cultivate.

But it comes out of God's love, not our need.

Yes, that is the good news, and sabbath is the gospel practice that teaches us that truth, body and soul, week after week after week.

Place habits before love, and you will be full of legalism, but place love before habits, and you will be full of the gospel. God's love for us really can change the way we live, but the way we live will never change God's love for us.

WEEKLY HABIT 4
SABBATH

THE HABIT AT A GLANCE

The weekly practice of sabbath teaches us that God sustains the world and that we don't. To make a countercultural embrace of our limitations, we stop our usual work for one day of rest. Sabbath is a gospel practice because it reminds us that the world doesn't hang on what we can accomplish, but rather on what God has accomplished for us.

THREE WAYS TO START

Pick a twenty-four-hour period. Ideally, we all sabbath on the same day. But it's better to do something than nothing. A pastor, a med student, a retiree, and a working mother would probably all choose different times to sabbath. I find Saturday sundown to Sunday sundown to be the best for my job and most of my friends' lives. The important part is picking a period and communicating it to the people who need to know or who are doing it with you.

Doing and not doing. You may need time to figure out what makes a worshipful and restful Sabbath for you. If you're just starting, it may help to write down three things to do and three things you want to avoid. They may change as you go, but writing them down will help you to not only think it through but also to be accountable.

Communal sabbaths. Sabbathing in community is a great way to get into a rhythm. Honor the sabbath in a friend group by having a regular communal meal. To keep it restful, you need a good division of labor. The host can't do all the cooking and all the cleaning. Maybe you can get in the habit of all bringing something to one house and all cleaning together afterward, with lots of conversation in between.

THREE CONSIDERATIONS

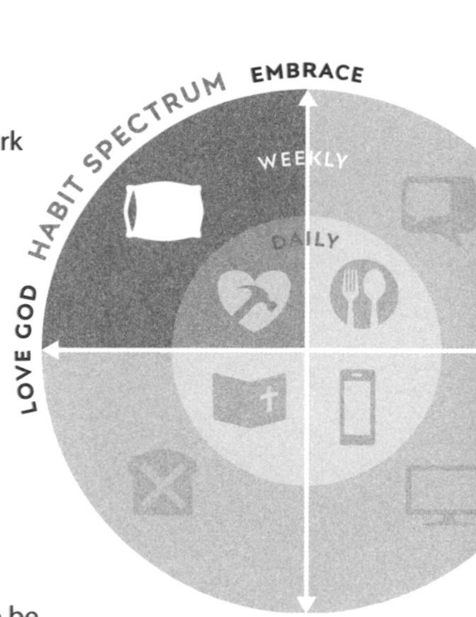

Work on the weekends. If you need to work a couple of hours or all day on Saturday to be able to take off Sunday, or vice versa, I recommend doing it. My family life is often better when I spend Saturday afternoon at the office so I can take all of Sunday off. This is much different than working a couple of hours both days. It's worth focusing and finishing so you can then focus on rest and realize that, in Jesus, all is finished.

Gentle away messages. You don't have to be awkward and in-your-face when you let people know you're having off time. A simple and nonjudgmental "I will be away from my computer until 8 p.m. today" email is elegant and goes a long way.

Electronic sabbaths. The most important way you might sabbath is to turn off screens. Yet this is not a law. I usually don't touch my computer, but in certain seasons, watching a long baseball game with my family is the heart of sabbathing for me. I recommend doing a month of sabbaths with no screens at all to get a real sense of what it's like. Then if you find that watching Netflix is your essence of sabbath, you can build that back in. But cutting the cord for a while gives you a taste of a weekly way of life that you may not know existed.

SABBATH IS THE ESSENCE OF OUR SALVATION. WE CAN REST BECAUSE GOD HAS DONE ALL THAT NEEDS TO BE DONE.

READING AND RESOURCES

The Tech-Wise Family, Andy Crouch
Sacred Rhythms, Ruth Haley Barton

EPILOGUE

ON FAILURE AND BEAUTY

To live intensely is a basic human desire
and an artistic necessity.
MICHAEL KIMMELMAN

"She has done a beautiful thing to me. . . . Truly, I say to you, wherever this gospel is proclaimed in the whole world, what she has done will also be told in memory of her."
JESUS (MATTHEW 26:10, 13)

Not the intense moment
Isolated, with no before and after,
But a lifetime burning in every moment.
T. S. ELIOT

THE HABIT OF FAILURE

Before dawn I woke slowly to the memory of the night before. A spontaneous happy hour with a client. A spring evening on the front porch. Getting carried away in conversation with a friend until one in the morning. *I should've been more responsible,* I thought as I sat up to the headache that comes with barely five hours of sleep and a beer after midnight.

My son Coulter was three months old at the time. I could hear him beginning to stir in the next room. *He shouldn't be up this early,* I complained to myself. I went and picked him up from the crib, and after taking a cursory kneel for morning prayer by the changing table, I began rocking him in the hopes he'd go back to sleep, which is laughable, because babies never just "go back to sleep."

I was frustrated because I was leading a group of people through the Common Rule habits, and I had just failed on many of my own goals. I was planning on fasting from alcohol—just blew that one. I was planning on being up earlier than my kids—there it goes. I had an email to write in order to check in and ask how everyone's habits were going, but I felt like an enormous hypocrite. I can never seem to do what I want to do—ever.

In my anger, I did what any normal human being does in such a circumstance. I opened my phone to scroll.

I didn't know what I wanted to search for. I just wanted something to catch my eye to get me thinking about something else. My heart was empty, and it was looking for something, *anything*, to fill it up.

Scripture before phone came to my mind as I opened it, because it had become a habit, so something felt weird if I started the day on my phone. *This is so stupid,* I thought. *It's a stupid legalistic rule that doesn't even matter.*

Then I paused. I noticed I was feeling something familiar: it was the feeling of habit.

I knew this feeling because I'd had it many, *many* times before. Usually it was associated with a bad habit such as having that late night snack, or saying something I knew was mean because it felt so good, or opening a browser window even though I was supposed to be working.

This time, in a strange twist, I was aware of a good habit—which I *really* didn't want to do. And I was trying to talk myself out of it, but I couldn't.

Fine, I'll do it. Ugh.

I shifted Coulter—who was chugging away at his pacifier and watching me wide-eyed, as if he could see the demon and the angel battling on my shoulders and was wondering how this one was going to play out. I opened a daily prayer app. Psalm 27 came up, and I began speed-reading it—like my eyes were sprinting through it. It was an absolute insult to the text.

Then it was as if someone had laid out a trip wire. I was running full speed through this psalm when I was suddenly cut off at the knees and went sprawling face first onto this line: "One thing I will seek after, to behold the beauty of the Lord" (Psalm 27:4, paraphrase).

I stopped. I read it again. Then I read it ten more times.

Suddenly everything had changed. I was standing next to a crib at the grey break of dawn, thinking about beholding beauty. I was thinking about the way something in me settles whenever I sneak into my kids' room and watch them sleep. I was thinking about the way my wife squints her eyes when she turns a page in a book. I was thinking of the way the sun splashes off the rocks in the James River every evening.

I was thinking about how God has filled my life with beautiful things. How everything inside of me was made to behold something beautiful. And how my destiny is to see the face of God looking back at me.

It was as if—in the words of my favorite Seamus Heaney poem—this line of the psalm had "come at me sideways / And caught my heart off guard and blew it open."[1]

The world was beginning to unfold, beginning with the knot that had been in me since I woke. I had spent my morning gazing at my tangled desires, but now my eyes were drawn up to the beauty of God and all that he has made.

Sin means that my heart curves inward, but the words of Scripture had cracked me open. And these pesky habits had set it all in motion again.

I did write my email that morning. I wrote it one-handed, on my phone, on the porch while holding Coulter. It began with this: "I woke this morning feeling like a failure and a fraud."

No other email I've sent has brought in so many grateful responses. Apparently nothing else I ever said about the Common Rule habits was as helpful as talking about failure—because failure is where we live.

This was the morning I realized that failure is not the enemy of formation; it is the liturgy of formation. How we deal with failure says volumes about who we *really* believe we are. Who we *really* believe God is. When we trip on failure, do we fall into ourselves? Or do we fall into grace?

Failure is the path; beauty is the destination. We walk toward beauty on the path of failure. Which is to say that formation occurs at the interplay of failure and beauty. No habits can be pursued for the purpose of success or productivity or a new and better you. They *must* be done for the vision of beauty. If the goal is self-help, failure will destroy you. But if the goal is beauty, failure makes that goal shine all the more brightly. So you get up and keep walking.

CURATING A BEAUTIFUL LIFE

Living a single, coherent life is one of the greatest dreams of human beings. In this sense a life of integrity is not about moral performance but simply about the pleasure of becoming *one* person. We long to be an integer, to be whole; instead, we are fractions of contradictory selves.

The art critic Michael Kimmelman has said that the greatest work of art is perhaps a life itself—curating all of life as a single witness to something grand. The best works of art are never single paintings or songs. They are the lives of great artists themselves, because great artists constrain their lives in order to produce the vision of beauty that has captured them.

I believe that paying attention to the work of habit is similar. It is best thought of as giving attention to the art of habit. It isn't about trying to live right; it's about curating a life. It is the art of living beautifully.

TURNING DAYS INTO LIVES

This is the vision of the apostle Paul in his letter to the Romans. After building his majestic argument for grace alone as the power of God's salvation, he turned to the reader to respond: "By the mercies of God . . . present your bodies as a living sacrifice, holy and acceptable to God" (Romans 12:1). In response to the work of God, give your whole body. Your whole life. After all, didn't Jesus present his whole body—his whole life—as a living sacrifice for us? "Imitate him." This is the idea. Paul went on to say, "Do not be conformed to this world, but be transformed" (Romans 12:2). The common root is *formation*.

If the question is "How does a human being offer a whole life to God? How do we live coherently?" the answer is *formation*, a word that connotes process.

We celebrate and yearn to live like the people who seem to have been able to concentrate their lives around a singular vision. William Wilberforce. Martin Luther King Jr. Dorothy Day. Gandhi. They seemed to be able to focus everything on the important things.

This vision—of a whole and coherent life—is the goal of a life curated by habit.

Yet what we so often overlook in our abstract hunt for beautiful lives is the striking plainness of the moments that make up the days that make up our lives. What we often overlook in our heroes are the one million tiny (but *so carefully chosen)* habits that got them there. By overlooking them, we overlook the way the most ordinary habits of limitation create the most extraordinary lives of meaning.

Anyone who studies habit inevitably happens upon the same surprise: how ordinary and simple the building blocks of the most complex and beautiful things are. Dr. Seuss wrote *Green Eggs and Ham* on a bet with his publisher that he couldn't write a book with only fifty unique words. Leonardo da Vinci began painting tiny brush strokes on a piece of poplar wood in 1503. Fourteen years and hundreds of thousands of brushstrokes later, that piece of poplar was the *Mona Lisa*, a masterpiece that experts are still x-raying to figure out how he painted impossibly thin brush strokes.

Remarkable things are built from the smallest persistent actions or by subjecting yourself to limitations. Sometimes these limitations are not voluntary. The famous portrait artist Chuck Close, after being struck with paralysis midcareer, created mammoth, realistic portraits using a brush strapped to his wrist and making small, tortured movements. To me, this makes it all the more interesting. Some argue that if you compare his work before and after the injury, it only solidified his style. He was already seeking limitations for the sake of creating something extraordinary. His works hang in the finest museums in the world. Many of them are portraits of his friends, twenty-by-twenty-foot testaments to the beauty of friendship in limited lives.

The former editor in chief of *Elle* magazine, Jean-Dominique Bauby, after suffering a stroke that left him immobile except for one eyelid, wrote his incredible memoir *The Diving Bell and the Butterfly* by blinking one eye to a secretary who would read through the letters of the alphabet to spell words. He died two days after it was published.

The connection between the ordinary and the extraordinary is through very small habits. Small things build up to great works of art. Limits often pave the way for new kinds of beauty. Obsessive control over seemingly meaningless things—such as the number of words used in a children's book—create some of the most meaningful and enduring works for all ages.

This urge to uncover the beauty of the ordinary and to understand the wide capacity of narrow limitations is an ethic profoundly lost in the technological frenzy of our modern world. But this remains stubbornly true: We will never build lives of love out of anything except ordinary days—simple, extraordinarily beautiful, but still ordinary days like those Robert Hayden celebrates in his famous ode to dutiful parenting, *Those Winter Sundays*. After describing the faithful routine of his father rising in the predawn cold to start a Sunday fire for the family, he concluded, "What did I know? What did I know / of love's austere and lonely offices?"

FAILING BEAUTIFULLY

Any process of curating a beautiful life will be laced with failure. That's what process means: learning as you go. But that's not an impediment to a beautiful life; it's *the way to it*.

My best friend Steve and I used to talk about what it means to become great, and we thought it meant focusing on how you handle success. Then life broke us down—as it will anyone. Now we talk about how any life is characterized much more by its failures than its successes. We believe that a great life comes not by the way you avoid failure, but by the way you handle failure.

Now, almost every week, we talk about *kintsugi* pottery. Kintsugi is the Japanese art of repairing a broken bowl by inlaying gold or other precious metals. The new bowl is stronger than the old one. The scars are the design. Your attention is drawn to the cracks and how they are mended. That is what you're supposed to see. The beauty is in the brokenness.

For those who focus inward—which is the legalistic gaze—failure destroys them. But look outward. Look for beauty, and you'll see that failure is making *you* the work of art. You are God's pot, with cracks inlaid with the gold of grace. You are more beautiful now because of the fault lines.

I am such a broken vessel. My life is riddled with error. My friends know it. Lord knows my family knows it. I ignore my wife. I curse angrily. I overpromise at work. I indulge in tobacco too often. I drink too greedily. I have extended daydreams in which I do absurdly amazing things and everyone loves me. I play those dreams out in arrogant comments I make out loud. I play them out in silent manipulations that no one can see—and worse, I'm good at it! I don't keep good enough track of my finances. I spend too much money on food. I don't give enough away. I turn my eyes from the injustices of systems I'm complicit in because they benefit me. I shut down my heart because I can't stand to hear another word of the world's pain that I can't fix. I throw myself at my career instead. I take on so much that I end up not having the time with my kids that I want. I yell at them when I'm mad.

My life is a sorry game of Whac-A-Mole. I use one habit to bang down one failure, and my soul rears up with a new one. There is no program that can simplify the fractions of my heart. There's no rule that can contain my chaos.

All of this is true *right now*. Right now as I'm writing about formational habits, it turns out I'm a mess of them.

People ask what my "life verse" is. Easy pick: "I find it to be a law that when I want to do right, evil lies close at hand" (Romans 7:21).

But here is the point. Look *at* me or *at* any other human being long enough, and you'll see nothing but a hypocrite. This will be true of every human being ever, no exceptions. But if you stand next to me and look where I'm looking, then we'll both see Jesus. He's the life we want. He's the life given for us. And the gold of the resurrection inlays all our fault lines. He is the one who lived the beautiful life. He is the one redeeming ours.

A beautiful life inspires a beautiful life. Even when the imitation of Christ is a sorry echo of the real thing, it's worth doing, because something worth doing is worth doing badly.

This is the ethic of the search for beauty, which is the only true worship. Like the woman pouring perfume on Jesus' feet, you lose yourself enough to embarrass yourself. You'll try anything, because you're lost in the one you love.

Those are the kind of habits worth cultivating—little habits of love, not carried out for success, not carried out to prove who you are, but cultivated because of a longing to love God and neighbor. That is a more beautiful life, one worth the constraints, one worth the failures.

Those are the kinds of habits that become the tiny strokes in the *Mona Lisa*, the shaking wrist painting a portrait of a dear friend. They become the days that become a life spent looking at the beautiful one, the one named Jesus, who with a glance can catch the heart off guard—and blow it open.

RESOURCES

HABITS AT A GLANCE

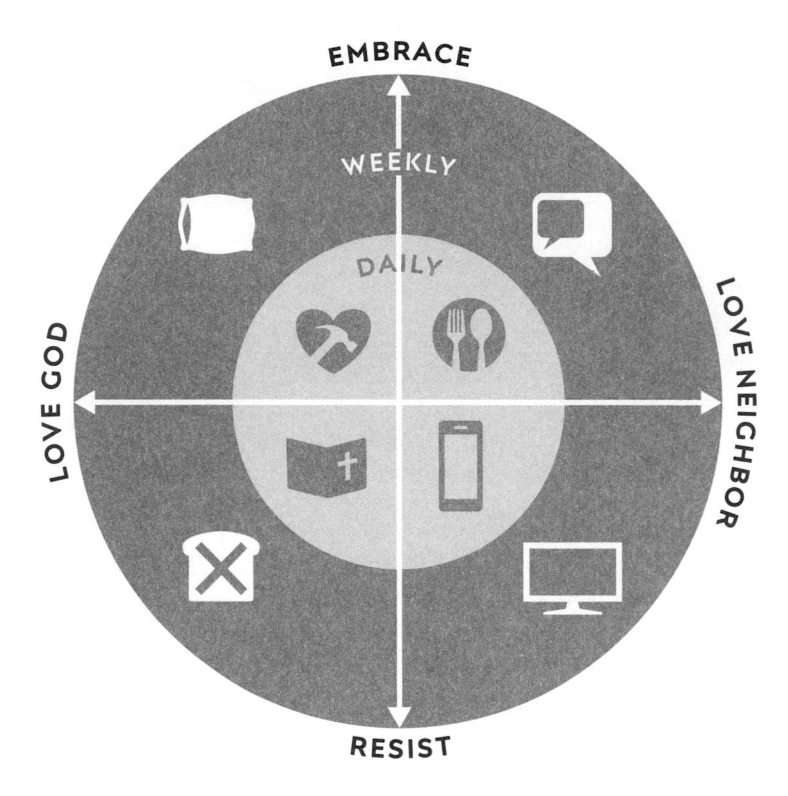

	DAILY HABITS		WEEKLY HABITS
1	Kneeling prayer three times a day	1	One hour of conversation with a friend
2	One meal with others	2	Curate media to four hours
3	One hour with phone off	3	Fast from something for twenty-four hours
4	Scripture before phone	4	Sabbath

HABITS IN A NUTSHELL

DAILY HABITS

Kneeling prayer three times a day

The world is made of words. Even small, repeated words have power. Regular, carefully placed prayer is one of the keystone habits of spiritual formation and is the beginning of building the trellis of habit. By framing our day in the words of prayer, we frame the day in love.

One meal with others

We were made to eat, so the table must be our center of gravity. The habit of making time for one communal meal each day forces us to reorient our schedules and our space around food and each other. The more the table becomes our center of gravity, the more it draws our neighbors into gospel community.

One hour with phone off

We were made for presence, but so often our phones are the cause of our absence. To be two places at a time is to be no place at all. Turning off our phone for an hour a day is a way to turn our gaze up to each other, whether that be children, co-workers, friends, or neighbors. Our habits of attention are habits of love. To resist absence is to love neighbor.

Scripture before phone

Refusing to check the phone until after reading a passage of Scripture is a way of replacing the question "What do I need to do today?" with a better one, "Who am I and who am I becoming?" We have no stable identity outside of Jesus. Daily immersion in the Scriptures resists the anxiety of emails, the anger of news, and the envy of social media. Instead it forms us daily in our true identity as children of the King, dearly loved.

WEEKLY HABITS

One hour of conversation with a friend

We were made for each other, and we can't become lovers of God and neighbor without intimate relationships where vulnerability is sustained across time. In habitual, face-to-face conversation with each other, we find a gospel practice; we are laid bare to each other and loved anyway.

Curate media to four hours

Stories matter so much that we must handle them with the utmost care. Resisting the constant stream of addictive media with an hour limit means we are forced to curate what we watch. Curating stories means that we seek stories that uphold beauty, that teach us to love justice, and that turn us to community.

Fast from something for twenty-four hours

We constantly seek to fill our emptiness with food and other comforts. We ignore our soul and our neighbor's need by medicating with food and drink. Regular fasting exposes who we really are, reminds us how broken the world is, and draws our eyes to how Jesus is redeeming all things.

Sabbath

The weekly practice of sabbath teaches us that God sustains the world and that we don't. To make a countercultural embrace of our limitations, we stop our usual work for one day of rest. Sabbath is a gospel practice because it reminds us that the world doesn't hang on what we can accomplish, but rather on what God has accomplished for us.

TRYING ONE HABIT OF THE COMMON RULE

Many people find that experimenting with one keystone habit is the best way to begin trying the Common Rule. Read this book while experimenting with one habit. This will give you an idea of why the rule is worth trying. The best introduction habits to start with are below, each of which is a keystone habit—a habit that helps to change other habits.

DAILY HABITS

Kneeling prayer three times a day. Framing the day in prayer gets you praying instead of always talking about praying or wishing you were a person who prayed. It also gets you structuring your day and thinking about it. Better habits frequently come along with it, such as having a morning or evening routine, taking breaks midday, and so on.

Scripture before phone. This habit is intended to get you reading Scripture daily. Almost as important, it helps keep your phone out of your morning, especially when you first wake up. Try this for a couple of weeks while you read the book. It often is the gateway to other new habits.

WEEKLY HABITS

One hour of conversation with a friend. The benefits of friendship are innumerable and wonderful. Committing an hour every week, during which you meet with someone and talk, can bring you out of isolation or spark a new phase in a community.

Sabbath. Sabbath changes the way you think about your week, the way you think about your work, and the way you think about time in general. When time is one long monotonous list of days, we fall apart. When we punctuate our days, weeks, and seasons with rest, we thrive. Start by planning how you will sabbath, and then try it out.

TRYING A WEEK OF THE COMMON RULE

Below is a template for reading and trying each of the habits in the Common Rule in one week. There's no need to try all the habits at once. Just read a chapter a day, they each take about fifteen minutes, and experiment with that habit.

This plan assumes you will try this in a group. If you don't have a group, I strongly suggest finding at least one other person. It's hard to make new habits alone.

Table 2. Try a Week of the Common Rule

<table>
<tr><th>Day</th><th>Read</th><th>Habit to Try</th></tr>
<tr><td rowspan="2">1
(Pick the day your group meets.)</td><td>Morning: Read the Introduction and Daily Habit 1.</td><td>Try kneeling prayer today.</td></tr>
<tr><td colspan="2">Evening: Meet with your group to discuss the idea of the Common Rule.</td></tr>
<tr><td>2</td><td>Read Daily Habit 2.</td><td>Try having an intentional, communal meal today.</td></tr>
<tr><td>3</td><td>Read Daily Habit 3.</td><td>Pick one hour to turn your phone off.</td></tr>
<tr><td>4</td><td>Read Daily Habit 4.</td><td>Before you use your phone today, read Scripture.</td></tr>
<tr><td>5</td><td>Read Weekly Habit 1.</td><td>Have an intentional one-on-one conversation with someone.</td></tr>
<tr><td>6</td><td>Read Weekly Habit 2.</td><td>Estimate the number of media hours you've watched this week. If it's already more than four, try not watching anything.</td></tr>
<tr><td>7</td><td>Read Weekly Habit 3.</td><td>Fast from something today.</td></tr>
<tr><td rowspan="2">8
(End on the day your group meets again.)</td><td>Morning: Read Weekly Habit 4 and Epilogue.</td><td>Make today a sabbath or write out some ideas to plan your next sabbath.</td></tr>
<tr><td colspan="2">Evening: Meet with group to discuss how the week went.</td></tr>
</table>

TRYING A MONTH OF THE COMMON RULE

When trying out a month of the Common Rule with a group, it's really helpful for all to see how the month is going to go. Below is a sample month and a daily template, so group members can see how they can arrange their days. Use this template so each person can customize to the month you're doing it in. You can also download templates from the website, thecommonrule.org.

DAILY HABITS

- Kneeling prayer at morning, midday, and bedtime.
- Pick Scripture readings: Psalms 1-30, Matthew 1-28, Romans; half of a chapter daily.
- The common meal will be__________with__________.
- Phone-off hour will be from__________to__________.

WEEKLY HABITS

Sun	Mon	Tue	Wed	Thu	Fri	Sat
1	2 Kick off month with an evening of discussion together.	3	4 Start the fast at 6 p.m. by skipping dinner.	5 Break the fast with a friend and have your weekly hour of conversation.	6	7 Count your media hours. (Don't beat yourself up. Just count to know.)
8 Sabbath	9 Meet to check in and discuss failures and realizations. Pray.	10	11 Start the fast at 6 p.m. by skipping dinner.	12 Break the fast with a friend and have your weekly hour of conversation.	13	14 Count your media hours.
15 Sabbath	16 Meet to check in and discuss failures and realizations. Pray.	17	18 Start the fast at 6 p.m. by skipping dinner.	19 Break the fast with a friend and have your weekly hour of conversation.	20	21 Count your media hours.
22 Sabbath	23 Meet to check in and discuss failures and realizations. Pray.	24	25 Start the fast at 6 p.m. by skipping dinner.	26 Break the fast with a friend and have your weekly hour of conversation.	27	28 Count your media hours.
29 Sabbath	30 Meet over a feast! Discuss what you learned.	31				

THE COMMON RULE FOR CONGREGATIONS

The Common Rule is supposed to be practiced in communities. That's where the "common" part comes from. When announcing a plan to practice the Common Rule together, make use of the resources at thecommonrule.org and do the following:

1. Create small groups. It's best for people to be divided into groups when trying habits, whether small groups, parishes, or accountability groups. Make sure these groups are small enough communities that people can talk about their experiences. As group members realize they have invisible bad habits, it's very important to be able to process failure and difficulties with others.

2. Cast vision for practicing the habits. People have to understand the *why*, otherwise these habits won't make sense. Most people haven't thought of this before, so don't just talk about all the things you're going to stop doing, which will come across as behavior management. Half of your people will be turned off, and the other half will have crushed souls. Remind them that our common dream is to change. We all want to change something. But usually we don't think about how much our small daily and weekly habits are getting in the way of becoming that new person. Consider doing a sermon series and/or pointing them to a video on the Common Rule website to get them thinking and motivated.

3. Emphasize the voluntary nature. If participants have the book, they can feel free to read up on one habit and experiment with it. If they're willing, suggest they try to follow along and practice some or all of the habits. It helps to have a core group—whether elders, small group leaders, or a group that's simply excited about a habit—lead the charge by committing to do it for the duration and guiding others. Note that formational habits can't and must not be forced. People have to be romanced into them. This often comes when they see or read how others are changed by the habits.

4. Set a time frame. A month works well for getting started. January is a good month to test new habits, but June typically isn't. A church season such as Lent or Advent is also a great opportunity to start. A week is usually too short for a congregation. Whatever you do, make sure people know how long they'll be trying out habits.

RESOURCES TO USE WITH GROUPS

This book. The most effective way for people to be drawn in is to have the opportunity to read the book.

The website. Make sure people have the thecommonrule.org web address so they can explore the habits on their own. Send it out in a newsletter or mailing, because people will forget the name.

Talks. Check the website for recent talks on the Common Rule. People are often interested but slightly confused when they first hear about a "rule of life" and the idea of committing to a set of habits. That's understandable. But after they listen to a talk, I often hear those same people say, "Okay, I need that." Consider having small groups or Sunday school classes watch a talk together.

Visualizing a month. When trying out a month of the Common Rule with a group, it's helpful for all to see how the month is going to go. Use the template on page 176. You can also download a template from the website so you can customize it to the month you're doing it in.

A PRAYER FOR THOSE TRYING THE COMMON RULE HABITS

If you're leading a group through the habits, pray this regularly with the participants. If everyone is comfortable praying it together, do a call-and-response version in which the leader reads the italicized sections, and the group reads the "may" responses in bold.

Father in heaven, who made us, who came to find us, who is redeeming us, we praise you! You took on the limitations of being human, that we might become the children of God.

May we see that our freedom comes only in service to you.

So Lord bless our days.

May our prayers be good work, and may our work become good prayers.

May our tables be full, and may they draw our neighbors in.

May we embrace good limits with technology and discover the beauty of being present with each other.

May we frame our days in the love letter of the Scriptures.

And Lord bless our week.

May our friendships be a fire around which all the lonely can gather.

May we choose stories that form us into people who seek beauty, who love justice, and who find each other.

May we discover in fasting that our emptiness is where we find your fullness.

May we sabbath in the soul rest of knowing that, in Jesus, "It is finished."

Lord, give us strength for all of this. We love only because you first loved us. When we stumble on our failure, may we fall into your grace.

May our lives become lights in the darkness, so that all may see your beauty. Amen.

THE COMMON RULE AS A RULE OF LIFE

While I encourage people to try the habits for at least a month, I actually do generally live according to them. There is much failure and constant adjustment, but the habits are a background rhythm of life that I come back to again and again. You may find your management team, family, church staff, or friend group wants to commit to them as a way of life. If that's the case, welcome! You may consider adapting a habit or two to your situation. You can download templates on the website that help you do that.

Here are some tips on adjusting the Common Rule as a long-term way of living:

Daily prayers: Consider using prayers written specifically for your community or prayers from a common liturgical calendar.

Daily meals: In a family this might be a way of life as family dinner. In an office or team setting this might be a standing lunch hour.

Phone off: This might become rhythms of silence or presence. Perhaps it's one hour your phone is off every day, or perhaps it's a daily rhythm of meditation, or a monthly rhythm of spending one day in silence. It could also be one hour as a family you spend together, undistracted.

Scripture before phone: Consider using a year-long reading plan or some other way to make sure you are formed in Scripture before you are formed in all the other competing narratives the world has to offer.

Sabbath: Week to week this may not change much. But sabbath rhythms should give rise to a way of life where you take rest as seriously as you take work.

Friendship: Vulnerable conversations every week may become a standing dinner or gathering between friends, it could culminate in an annual retreat or some other way that you keep the heartbeat of friendship echoing through all of life.

Curating media: Usually curating a limit every week turns into a new way engaging with media. The goal of this habit is not to count hours but to become a person who thinks about media differently and curates as a way of life.

Fast: Fasting as a way of life is about bodily restraint. This might often interplay with an exercise routine or regular times or seasons away from certain food or drink.

THE COMMON RULE FOR DIFFERENT WALKS OF LIFE

FOR SKEPTICS

Someone may have handed you this book and said you'd like it even though you won't agree with some of it. You may be someone who used to follow Jesus but then something happened. Maybe you're someone who sympathizes, but you aren't ready to call Jesus by name. Maybe you think spiritual rhythms are cool and useful, no matter what you believe. Maybe you're just looking for who God is. If so, I think these will help you. And I pray you'll find what you're looking for—or better yet, that he'll find you.

The Common Rule for Skeptics

Daily Habit	Variation / Focus
3x Prayer	Being silent at regular intervals is an amazing way to punctuate your day, and silence is the beginning of all prayer. Begin your day with a couple of minutes of silence, repeat midday at work, and end the day with the same. After a couple of weeks, begin one of those sessions with asking something of God. Ask anything, no matter how honest. He can take it. Then be quiet and listen.
Meal with Others	No matter who you are, you need people. Regular communal meals are not just a way to nourish your body—they are a way to nourish something your soul longs for.
Hour with Phone Off	No one really knows who they are until they are quiet. Use an hour with your phone off every day to create space to reflect. This can help prevent confusing yourself and what you believe by the constant noise of the phone.
Scripture Before Phone	For you this may be just reading before phone. There is power in the written page. If you're looking to examine Christianity, try the Gospel of John, C. S. Lewis's *Mere Christianity*, Tim Keller's *The Reason for God*, or Rick Warren's *The Purpose Driven Life*. If you're not ready or comfortable reading the Bible or a book about the Bible, just start your day with poetry or a any writer who challenges you and expands your mind.

Weekly Habit	Variation / Focus
Sabbath	Studies show that productivity drops off sharply after fifty hours of work a week. Don't waste your time or anyone else's trying to prove how much you can work. Become a whole person by taking time off and doing something that refills you.
Hour of Conversation	If you're trying to understand Christianity, meet regularly with a follower of Jesus you trust. But remember that we are the best and worst reflections of him. We're not good enough, so don't expect much more than a mess just like you. An hour of conversation a week is a great way to pursue friendship, which we all need.
Curate Media to Four Hours	Like it or not, you're not as independent of a thinker as you wish you were. You become the stories you watch. Choose them carefully. Choose ones that challenge your perspective, ones that you know you won't agree with.
Fast	Restraint is healthy, no matter who you are. I know of no one without vices. You may choose to fast as a spiritual search. Alternatively, you may choose to abstain regularly from something you know has a grip on you. It may help to know that in the Christian understanding of fasting, a fast is not for self-betterment but for seeing past yourself to the One who sustains you. In essence, it's not about you.

FOR PARENTS

Full-time parenting may be one of the most difficult careers out there. If we become our habits, and our kids become us, then our kids become our habits. For parents, attentiveness to your own habits is the beginning of teaching your kids how to live—not to mention how to use technology wisely.

The Common Rule for Parents

Daily Habit	Variation/Focus
3x Prayer	Writing a short prayer for your children with your spouse is a good way for both of you to focus on praying for the same things for your kids. Try a couple of sentences and pray them three times a day.

Meal with Others	Cultivating a habit of a family breakfast or family dinner is an amazing way to grow together. Pick which meal is best for you—and don't imagine it's going to be easy. It will be messy and loud, and there will be tons of prep and cleanup. Don't allow electronic devices at the table, and no one leaves until excused. This is hard, but all good things are. The table is where you learn to know and love each other.
Hour with Phone Off	The phone-off hour should be paired with engaging with your kids—whether it's wrestling, building, dressing up, talking, or board games. An undistracted hour of engaging with them is worth its weight in gold.
Scripture Before Phone	I don't know how to parent outside of being parented by Jesus. It's useful to my body and soul to be in the Word—however briefly—before my kids wake up. My wife is not like this. She prefers to read the Bible in the afternoon. But in no case does using a smartphone first thing in the morning set up parents for a healthy day. Try to get it out of your morning routine.
Weekly Habit	**Variation/Focus**
Sabbath	If the non-lead parent during weekdays can carry the brunt of the work on sabbaths, that's ideal. But not everyone can. If not, consider trying a month of communal gatherings with other friends or family where the load is lighter. This is often specific to each family, so you'll need to practice a lot together and measure success in the long run, not the short term.
Hour of Conversation	Adult conversation is a precious commodity to a full-time parent. This should happen when the kids aren't around. Talking at the park while the kids play just doesn't cut it. Sometimes it's great to have another parent to talk to, but I also find it helpful to talk to someone who is not in the same parenting stage.
Curate Media to Four Hours	Watch something together as a family—that is, make your watching communal. Having a canon of shared movies or TV shows is an incredible family bonding mechanism. Also, I find that even the most "appropriate" movies require explanation for my children. Discuss where you are seeing virtue worth imitating and where there is vice worth avoiding.
Fast	Often a twenty-four-hour fast from food is inadvisable for parents—and especially for nursing or pregnant mothers. Fasting from sugar or dessert works well. Sometimes fasting from social media or sports does too. The goal is to pick something that will make you feel a lack and to lean into Christ's abundance when you do.

FOR THE WORKPLACE

The variations below assume that you're pairing the Common Rule practices with an intense project or a season at work in order to keep healthy limits and rhythms as you focus and work hard. It's ideal to have someone else in your office who joins you in the practices.

The Common Rule for the Workplace

Daily Habit	Variation / Focus
3x Prayer	The midday prayer becomes especially important during trying seasons at work. The goal is to frame work in love, not to let work frame you in stress. Consider a short midday meet-up with a coworker to pray.
Meal with Others	One of the most helpful things for me has been having a coworker or two with whom I have standing coffee breaks or standing lunches. And I also protect family meals by not letting work encroach on family dinner unless it is a true emergency. This depends on your life stage, so do what works for you.
Hour with Phone Off	Work is exhausting, so having your phone off for a while after work is a good way to have a mini-sabbath. It's also important to have space at work to focus. Consider having a box at the office door where you and others can drop off your phones. Offer charger cords to incentivize.
Scripture Before Phone	Pair Scripture reading with a wake-up hour to get into a disciplined schedule for work. Usually my best work comes in the morning before anyone else is up. If that's true for you, anchoring your morning in Scripture and then in quiet work can create a powerful and sustainable morning routine.
Weekly Habit	**Variation / Focus**
Sabbath	Rest will be the most important thing that happens to your career. Rest means getting enough sleep on a daily basis and also having a sabbath on a weekly basis. Good work is from God and for neighbor, and you can't do it right without worshiping corporately and resting regularly.
Hour of Conversation	One of the most dangerous consequences of tough seasons of work is that after a few, we tend to grow distant from our friends. Yet consider the impact of carving one hour out of your week. It's a low cost to work and a huge impact to life. That hour can change your whole life. If we can't make that simple decision for our own life, it's a wonder why we trust ourselves to make management decisions that affect other people's lives.

Curate Media to Four Hours	Consider making media time only communal, so you're pushed into community during relaxation times. Also, binge watching is often a coping mechanism for stress. Curating your media intake to four hours may actually make you more productive because you're putting the guardrails up.
Fast	Fasting is countercultural, especially at work. If you have a job where fasting would decimate your energy or impair your ability to do your job, pick something else to fast from. I find that fasting is a great way to reset my gaze from people-pleasing (which is tempting at work) to trusting that God will help me do good work.

FOR ARTISTS AND CREATIVES

Creativity is not something that happens randomly. There is no muse. It is a lie. Creativity happens when you ruthlessly remove distractions. Art is the product of stringing together small bits of work, day after day after day. For example, practicing the Common Rule while writing this book was especially important to my process.

The Common Rule for Artists and Creatives

Daily Habit	Variation / Focus
3x Prayer	Prayers for inspiration and focus may be helpful. When I begin a project, I write a few prayers at the beginning—when I am clear about what really matters—and then pray them through the complications that inevitably come. I did that for this book project.
Meal with Others	Consider a meal at the end of the day with family or a roommate as a way to end work. We often don't know how to call it quits, and coming home to a community that welcomes you is one of the best ways to punctuate the end of a workday.
Hour with Phone Off	In every new project, the beginning phase is intensely creative and requires hours of undistracted time. Then comes the edits and the reworking, which is hard but less soul work. In the creative phase, consider flipping this to one hour with your phone on and have it off the rest of the time. Often creativity comes simply from removing distractions.

Scripture Before Phone	Consider having a carefully planned morning routine where you get up at the same time, read a passage, eat what works for you, and move right on to your creative work. Morning liturgies/routines are really important for most creatives.
Weekly Habit	**Variation / Focus**
Sabbath	It's especially important not to work on your creative endeavor during sabbath because it's your work. Your soul needs a break from it.
Hour of Conversation	Seek out those who understand your art or creative process or who are working on their own. Isolation is a great danger for the artist, and friendship is often what saves us from ourselves.
Curate Media to Four Hours	Consider curating for inspiration and avoiding distractions. Make a list of the documentaries, podcasts, speeches, or lectures that get you thinking about the way other artists see and indwell their craft. Watch or read about the great artists in your craft or canon. Indwell their work so you can riff on it, and avoid media that numbs your mind.
Fast	Pair fasting a diet or exercise plan so your body is keeping up with your mind. When the mind gets ahead of the body, there's always a crash. You have no work outside of a functioning body.

FOR ENTREPRENEURS

Whether you like it or not, when you're building an organization, you're replicating yourself—the best and the worst of your qualities. To be attentive to who you are becoming is to be attentive to who your employees are becoming and what your organization is becoming. If you're trying the Common Rule as an entrepreneur, absolutely do it with your cofounders or maybe even with your senior management team. If not everyone follows Jesus, they may consider adapting the following Common Rule for Skeptics on page 183.

The Common Rule for Entrepreneurs

Daily Habit	Variation / Focus
3x Prayer	Anyone at the helm of a new organization feels the crush. There are oceans more to do than can actually be done. Days can mean chaos, with you putting out fire after fire. That's why it's all the more important to punctuate your day, so that you frame the chaos instead of letting the chaos frame you.
Meal with Others	If you have a family, a regular family meal could be a very important way to be consistently present, even if you're in a crazy stage at work. If you do not, lunches could be ideal—a regular communal break for you and your team.
Hour with Phone Off	You can't be omnipresent, even though it seems like everyone needs you to be. What they actually need is to know is the precise hour every day you're not available. This will help them to figure out how to move things along without you, which is the end goal of any organization.
Scripture Before Phone	You can't just do output as an entrepreneur; you must have constant meaningful input. Consider a morning routine that blocks out time to read the Bible first and then whatever kind of book expands your imagination and vision for your company and your team. Don't start the day in the infinite tasks you have. They will take over everything, they will cloud your ability to have vision—and that is your most important job.
Weekly Habit	**Variation / Focus**
Sabbath	This is the one thing you think you cannot do, but it is the one thing you must do. Consider how important regular and meaningful time off is for you and for your employees. Think about ways to structure your organization to make sure everyone has the option to—and is even encouraged to—disengage one day of their choice each week. Make sure that your habit of taking off one day a week doesn't require someone else to work seven days.
Hour of Conversation	For an entrepreneur, having a regular hour of conversation with a mentor may be the wisest choice here. It should be someone who isn't in your stage and can reflect back on it. It's unlikely that a mentor could meet every week, so seek a regular hour each week with friends to make sure you're attending to all of life, not just your work.

Curate Media to Four Hours	Entrepreneurs can get tunnel vision when they spend all their time thinking about their organization. Consider curating your media to remind you of those in the world who are not like you, those who are not your potential customers, those who need charity. This may inspire your organization to tithe revenue to a certain cause or provide space for your employees to do volunteer work.
Fast	Use a fast to protect your health from neglect or excess. An exercise routine paired with one day a week when you fast from food or fast from a guilty pleasure can keep your body up to the long race of entrepreneurship.

FOR ADDICTS

I'm prone to excess. I struggle mightily with moderation. In earlier stages of my life I've been functionally dependent (if not chemically addicted) to alcohol or pornography. Here's the amazing news: God changed me. Those times are long gone. I'm not who I used to be, and you aren't condemned to be who you are now. Grace is real, which means change is too. If you are struggling with or recovering from addiction to alcohol, prescription drugs, sex, or something else, you probably know that this process can't be simply removing the addiction; *you have to fall in love with something else*. Living according to the Common Rule with a companion who is helping you in the process can help you focus on forming a new life without the addiction that rules you. You may be a slave to your drug, but there is freedom in new limitations. There is a better master. His name is Jesus.

The Common Rule for Addicts

Daily Habit	Variation / Focus
3x Prayer	Prayer changes reality. You will not become a new person outside of prayer. Practice being silent, practice pleading with God over and over, practice framing your day in Jesus' love for sinners. You can't be shamed out of an addiction; you can only be loved out of it.

Meal with Others	Addictions are often paired with hiding from community or running to the wrong community. You need to regularly be with the people who are helping you out. Use regular meals to engage those people. Let them in; don't shut them out.
Hour with Phone Off	Consider carefully whether stress is driving your addiction. For me and most of my friends who have addiction struggles, there may be something you feel like you have to do to keep the world from coming apart. If that's work, keeping your phone off may help. If it's something else, think about how you can remove yourself from that stress at a regular time each day.
Scripture Before Phone	You need to know that you are loved no matter how much you fail. And you need to know that's not a platitude; those are Jesus' words to sinners. Cling to them each morning. You aren't the worst thing you've done, even if that thing happened again last night.

Weekly Habit	Variation / Focus
Sabbath	You're probably used to resting by finding release in acting out your addiction. You have to find new ways, and they will likely be only approximations of the old drug. Seek them out in nature, in others, in sports, in hobbies, and in exercise. Whatever it is that unwinds you, run to that rest. God is there.
Hour of Conversation	If there is one keystone habit for an addict, this is it. Someone—just one person—must know everything about your addiction and your relapses. Lying and hiding is the key to starting and continuing an addiction. Truth-telling is the keystone habit out. Find someone you trust, and do the most terrifying and most freeing thing you'll ever do: Tell that person everything. Don't gloss over anything. You must be exposed completely before the light comes in.
Curate Media to Four Hours	Some media can be triggers for longing for the old thing. You may need to be without a computer, only watch things with others, or pick things that don't lead you to the old places. There are also things to run to, not just from. Run to stories of beauty and redemption. We who struggle with addiction are always going to need those stories.
Fast	You may feel like your whole life is a fast, but fasting from food regularly is a way to create an ongoing tone of moderation and restraint. It can help you see that God is more than your drug. It reminds us that we're actually happiest when we don't get what we want but when we get what Jesus wants. Let this truth work through your life by fasting. Pairing exercise routines with this can be a really effective way to train your body to love new things.

FOR THOSE DEALING WITH MENTAL ILLNESS

The Common Rule was born out of my struggle against a bizarre anxiety that seemed partly to be brought on by circumstances and partly by chemicals going wild in my mind and body. I'll probably never understand all that happened. What I do know is that formational habits changed everything for me. I can't promise it will be that way for you, however. While they may not make all your troubles disappear, the habits can make you stronger, more stable, more joyful, and better equipped to fight the battle that rages in your mind. God is with you, and your friends are too. Find them and practice these habits together. *Note: I am not an expert, just someone who can share what helped in my own journey so please, please be open with your friends and family and by all means go see a physician or a mental health professional about your own symptoms.*

The Common Rule for Those Dealing with Mental Illness

Daily Habit	Variation / Focus
3x Prayer	If you have anxiety, structure your days. If you're depressed, plan something to get you out of bed. I found that when my mind was malfunctioning, blank spaces and silence were terrifying. Try written, habitual, or liturgical prayers. Pray them until they pray you.
Meal with Others	Your mind is going to make you want to withdraw, but rhythms of healthy community are what you need. Food and diet often cause or exacerbate these phases. Be attentive to what you're eating as much as how you're eating with others. A carefully planned diet under supervision and eaten with others can be a strong aid to recovery.
Hour with Phone Off	Silence is terrifying to the troubled mind, but it is also like working out: it hurts bad, and then you get stronger. Turn your phone off to practice silence. Know that when your mind rages, your soul doesn't have to. Evil lurks in wilderness, but God is there too, and he always wins.

Scripture Before Phone	Mental illness—even when chemically caused—always involves telling yourself an untrue story about yourself. You desperately need Scripture to tell you the way the world actually is. Find a devotional you trust, or read through an uplifting book. Make this the way your day begins, and consider going back to it at the time of day your depression or anxiety usually is the worst.

Weekly Habit	Variation / Focus
Sabbath	Coping mechanisms abound for bouts with mental illness, but sabbath offers a way to engage rhythmically and healthily with the joy and peace you need. Worship heartily, even though you may not find understanding in your congregation. You don't need everyone to understand you. The Holy Spirit is there. Don't withdraw.
Hour of Conversation	Much of this struggle is about you being brought out of your isolation. You have to tell the truth of what's happening to other people, and they have to tell the truth back to you. You are not your scariest thoughts. For me, it took my friends constantly speaking truth to counter the lies in my mind. Find trusted friends, and talk to them each week. Also believe them. Don't believe the reels in your head. They will spiral you. Cultivate a healthy doubt of your own fears. The voices in your head can sink you, but the voices of community will lift you up.
Curate Media to Four Hours	Be careful with media. Often a nonstop stream of noise exacerbates anxiety. Consider shutting out everything except trusted voices and the sources that bring you out of navel gazing and draw your eyes to God and to the world that needs you.
Fast	When I was really battling anxiety, a day of fasting wasn't enough. Three-day fasts (which take practice) were some of the most healing and life-giving times I had. Have someone experienced mentoring you in this, and get an okay from your doctor or therapist. Know that when you get low on energy, your body will give you an initial barrage of odd emotions. If you have the strength, try to push through it; mental clarity is often right on the other side.

ACKNOWLEDGMENTS

I would like to acknowledge the following people for their influence on my life, and thus this book. My gratitude is owed:

To Lauren, I would rather curate life with no one else. To Matt and Steve for being friends like brothers. To Mark, Frank, Dan, and David for being brothers like friends. To Anne, Mary Alice, Rachel, and Mary Catherine for your sisterhood to Lauren and all the help with our boys in such a difficult season. To Mom for modeling how to live with a hunger for life. To Dad for all those mornings you woke me up to pray. To Fred and Teresa for your incredible generosity toward us, especially in all the babysitting that made writing retreats possible. To all of The Cast (including your wives) for showing me covenant friendship. To Drew for the many book recommendations and always telling me who I was paraphrasing. To Barrett for the cocktails and conversations on these subjects. To Jason for all the evenings of friendship in DC. To Andrew and Chris for your companionship in this season, and to Rob for the companionship in so many seasons. To the House of Pain for laying the foundations of Christian community. To Todd and Kevin for caring enough to disciple me. To Scott for the mentorship that expanded my view of work and the whole East Asia community for teaching me how to live in exile. To Corey for telling me what a rule of life was (and especially for the unmerited trust you've shown me and your generosity to me in the introduction to InterVarsity Press). To the Third Church community for suffering through hearing me talk about habits ad nauseam. To McGuireWoods for teaching me how to be a good lawyer, and to

Gammon and Grange for inviting me in. To RHG for all those books you gave me that you probably don't know changed my life. To John Hagadorn and Bill Wilson for teaching me to love to read. To Richmond Hill for the writing retreats and the sacred space where this book was conceived. To John Rood for introducing me to Mars Hill Audio. To Meaghan, Rachel, Dan, David, Steve, Mark, Drew, Matt, Christian, and Lauren for reading the many early, awful versions of this manuscript. To Al for being a gentle and wise editor. To David Fassett for the excellent design work on the cover. To the people whose influence on my thinking in these subjects has been so significant that it's impossible to properly footnote: Ken Myers, Greg Thompson, James K. A. Smith, Tish Harrison Warren, Ruth Haley Barton, and Tim Keller. To Andy Crouch for being so generous to me with your influence and encouragement on this project. And finally, to Whit for your leadership among the brothers—"I have reservations about so many things, but not about you." To Asher for your beautifully mischievous energy—you make spontaneous donut trips all the more fun. To Coulter for leaving me the two-foot strip of the bed where I realized that I was happier when living in limitations—I started this project with you in my arms. And to Shep for your warm company on the couch during proofreading sessions—I suspect you'll be a great friend. And to Jesus the King for many things, but especially for walking up to me that morning in late January when I felt I couldn't write another sentence—Maranatha.

NOTES

INTRODUCTION

[1]David Foster Wallace, "2005 Kenyon College Commencement Address," May 21, 2005, *Kenyon College Alumni Bulletin*, http://bulletin-archive.kenyon.edu/x4280.html.

[2]David Brooks, "The Machiavellian Temptation," the *New York Times*, March 1, 2012.

[3]William James, *Talks to Teachers on Psychology: And to Students on Some of Life's Ideals* (New York: Henry Hold and Company, 1914), 64.

[4]Charles Duhigg, *The Power of Habit* (New York: Random House, 2012).

[5]Annie Dillard, *The Writing Life* (New York: Harper & Row, 1989).

DAILY HABIT 1

[1]Annie Dillard, *The Writing Life* (New York: Harper & Row, 1989).

DAILY HABIT 2

[1]Christine D. Pohl, *Living into Community: Cultivating Practices That Sustain Us* (Grand Rapids: Eerdmans, 2012).

[2]Matthew Levering interview, *Mars Hill Audio Journal*, volume 135, part 2, 2013.

[3]By secular age, I have in mind the way Charles Taylor, and those who have interpreted him such as James K. A. Smith, mean when they use that phrase. See James K. A. Smith, *How (Not) to Be Secular: Reading Charles Taylor* (Grand Rapids: Eerdmans, 2014).

[4]I am following in Lesslie Newbigin's footsteps here; see Lesslie Newbigin, *Foolishness to the Greeks: Gospel and Western Culture* (Grand Rapids: Eerdmans, 1986) and Lesslie Newbigin, *The Gospel in a Pluralist Society* (Grand Rapids: Eerdmans, 1989).

[5]Madeleine L'Engle, *Walking on Water: Reflections on Faith and Art* (New York: Random House, 1980), 191.

[6]Don Everts and Doug Schaupp, *I Once Was Lost* (Downers Grove, IL: InterVarsity Press, 2008), 29.

DAILY HABIT 3

[1]Adrian F. Ward, Kristen Duke, Ayelet Gneezy, and Maarten W. Bos, "Brain Drain: The Mere Presence of One's Own Smartphone Reduces Available Cognitive Capacity," *Journal of the Association for Consumer Research* 2.2, April 2017, 140-54.

[2]See Cal Newport, *Deep Work* (New York: Grand Central Publishing, 2016).

[3]Kyle David Bennett, *Practices of Love: Spiritual Disciplines for the Life of the World* (Grand Rapids: Baker, 2017), 125.

WEEKLY HABIT 1

[1]C. S. Lewis, *The Four Loves* (New York: HarperCollins, 1960).

[2]Sherry Turkle, *Reclaiming Conversation: The Power of Talk in a Digital Age* (New York: Penguin Books, 2015), 143.

WEEKLY HABIT 2

[1]Stanley Hauerwas, *Hannah's Child: A Theologian's Memoir* (Grand Rapids: Eerdmans, 2010), 156.

WEEKLY HABIT 3

[1]Speech given by Dr. Martin Luther King Jr. at Stanford University on April 14, 1967, https://kinginstitute.stanford.edu/news/50-years-ago-martin-luther-king-jr-speaks-stanford-university.

[2]Michael Eric Taylor, "The African-American Community of Richmond, Virginia: 1950-1956," (1994), Master's Theses 1081, https://scholarship.richmond.edu/masters-theses/1081/.

[3]Ernest Becker, *The Denial of Death* (New York: Simon & Schuster, 1973).

WEEKLY HABIT 4

[1]I'm indebted to Tim Keller's teaching on sabbath for the idea of the rest beneath rest.

[2]This quip is informally attributed to Abraham Joshua Heschel but likely traces back to a common Jewish saying.

[3]Julian of Norwich, *Revelations of Divine Love*, Oxford World's Classics, trans. Barry Windeatt (Oxford: Oxford University Press, 2015), 20.

EPILOGUE

[1]Seamus Heaney, paraphrase of “Postscript,” *Seamus Heaney* (Cambridge: Harvard University Press, 1998).